NORTHERN EUROPE
An Environmental History

Other Titles in
ABC-CLIO'S
NATURE AND HUMAN SOCIETIES SERIES

The Mediterranean: An Environmental History,
J. Donald Hughes

Northeast and Midwest United States: An Environmental History,
John T. Cumbler

FORTHCOMING TITLES:

*Australia, New Zealand, and the Pacific:
An Environmental History,* Donald S. Garden

*Canada and Arctic North America:
An Environmental History,* Graeme Wynn

Russia: An Environmental History,
Alexei Karimov and Irina Merzliakova

South Asia: An Environmental History,
Richard Hugh Grove

Southeast Asia: An Environmental History,
Peter Boomgaard

Sub-Saharan Africa: An Environmental History,
Gregory H. Maddox

*United States Great Plains and Intermountain West:
An Environmental History,* James E. Sherow

United States South: An Environmental History,
Donald E. Davis

United States West Coast: An Environmental History,
Adam M. Sowards

NATURE AND HUMAN SOCIETIES

NORTHERN EUROPE
An Environmental History

Tamara L. Whited, Jens I. Engels,
Richard C. Hoffmann, Hilde Ibsen,
and Wybren Verstegen

Santa Barbara, California • Denver, Colorado • Oxford, England

Library of Congress Cataloging-in-Publication Data
Northern Europe : an environmental history / Tamara L. Whited . . . [et al.].
   p. cm.—(ABC-CLIO's nature and human societies series)
  Includes bibliographical references and index.
   ISBN 1-85109-374-5 (hardback : alk. paper)—ISBN 1-85109-432-6 (ebook)
1. Human ecology—Europe, Northern—History. 2. Europe, Northern—Environmental conditions. I. Whited, Tamara L. II. Series: Nature and human societies.

GF540.N67 2005
304.2'0948—dc22

                                                              2005017733

08 07 06 05 / 10 9 8 7 6 5 4 3 2 1

This book is also available on the World Wide Web as an eBook.
Visit abc-clio.com for details.

ABC-CLIO, Inc.
130 Cremona Drive, P.O. Box 1911
Santa Barbara, California 93116-1911

This book is printed on acid-free paper ∞.
Manufactured in the United States of America

CONTENTS

SERIES FOREWORD

Long ago, only time and the elements shaped the face of the earth, the black abysses of the oceans, and the winds and blue welkin of heaven. As continents floated on the mantle they collided and threw up mountains, or drifted apart and made seas. Volcanoes built mountains out of fiery material from deep within the earth. Mountains and rivers of ice ground and gorged. Winds and waters sculpted and razed. Erosion buffered and salted the seas. The concert of living things created and balanced the gases of the air and moderated the earth's temperature.

The world is very different now. From the moment our ancestors emerged from the southern forests and grasslands to follow the melting glaciers or to cross the seas, all has changed. Today the universal force transforming the earth, the seas, and the air is for the first time a single form of life: we humans. We shape the world, sometimes for our purposes and often by accident. Where forests once towered, fertile fields or barren deserts or crowded cities now lie. Where the sun once warmed the heather, forests now shade the land. One creature we exterminate only to bring another from across the globe to take its place. We pull down mountains and excavate craters and caverns, drain swamps and make lakes, divert, straighten, and stop rivers. From the highest winds to the deepest currents, the world teems with chemical concoctions only we can brew. Even the very climate warms from our activity.

And as we work our will upon the land, as we grasp the things around us to fashion into instruments of our survival, our social relations, and our creativity, we find in turn our lives and even our individual and collective destinies shaped and given direction by natural forces, some controlled, some uncontrolled, and some unleashed. What is more, uniquely among the creatures, we come to love the places we live in and know. For us, the world has always abounded with unseen life and manifest meaning. Invisible beings have hidden in springs, in mountains, in groves, in the quiet sky and in the thunder of the clouds, and in the deep waters. Places of beauty from magnificent mountains to small, winding brooks captured our imaginations and our affection. We have perceived a mind like our own, but greater, designing, creating, and guiding the universe around us.

The authors of the books in this series endeavor to tell the remarkable epic of the intertwined fates of humanity and the natural world. It is a story only now coming to be fully known. Although traditional historians told the drama of men and women of the past, for more than three decades now many have added the natural world as a third actor. Environmental history by that name emerged in the 1970s in the United States. Historians quickly took an interest and created a professional society, the American Society for Environmental History, and a professional journal, now called *Environmental History*. American environmental history flourished and attracted foreign scholars. By 1990 the international dimensions were clear; European scholars joined together to create the European Society for Environmental History in 2001, with its journal, *Environment and History*. A Latin American and Caribbean Society for Environmental History should not be far behind. With an abundant and growing literature of world environmental history now available, a true world environmental history can appear.

This series is organized geographically into regions determined as much as possible by environmental and ecological factors, and secondarily by historical and historiographical boundaries. Befitting the vast environmental historical literature on the United States, four volumes tell the stories of the North, the South, the Plains and Mountain West, and the Pacific Coast. Other volumes trace the environmental histories of Canada and Alaska, Latin America and the Caribbean, Northern Europe, the Mediterranean region, sub-Saharan Africa, Russia and the former Soviet Union, South Asia, Southeast Asia, East Asia, and Australia and Oceania. Authors from around the globe, experts in the various regions, have written the volumes, almost all of which are the first to convey the complete environmental history of their subjects. Each author has, as much as possible, written the twin stories of the human influence on the land and of the land's manifold influence on its human occupants. Every volume contains a narrative analysis of a region along with a body of reference material. This series constitutes the most complete environmental history of the globe ever assembled, chronicling the astonishing tragedies and triumphs of the human transformation of the earth.

Creating the series, recruiting the authors from around the world, and editing their manuscripts has been an immensely rewarding experience for me. I cannot thank the authors enough for all of their effort in realizing these volumes. I owe a great debt too to my editors at ABC-CLIO: Kevin Downing (now with Greenwood Publishing Group), who first approached me about the series; and Steven Danver, who has shepherded the volumes through delays and crises to publication. Their unfaltering support for and belief in the series were essential to its successful completion.

Mark Stoll
Lubbock, Texas

ACKNOWLEDGMENTS

Tamara L. Whited acknowledges the assistance of the following individuals and institutions: my home university, Indiana University of Pennsylvania, allowed me time away from my usual professional responsibilities so that I could devote myself to thinking and writing about the big picture of northern European environmental history. In particular I thank the Senate Fellowship Committee and the University-Wide Sabbatical Committee for awarding me release time and a sabbatical leave.

The success of this project depended in many ways on Mark Stoll's perseverance and encouragement: he helped put together our scholarly team, read countless drafts, and gave cogent advice on many fronts with his characteristic enthusiasm and patience. In writing the first two chapters I traveled far from my usual scholarly environment and owe a great deal to Daniel E. Lieberman and R. Scott Moore for correcting, pruning, and smoothing my initial efforts. The final two chapters benefited from readings by Wayne K. Bodle, Jean-Yves Boulard, and Angela Whited. I warmly thank my coauthors, Jens Ivo Engels, Richard Hoffmann, Hilde Ibsen, and Wybren Verstegen, for the opportunity to discover their talents and to work closely with them, from our early discussions of major themes to the final polishing. Wybren Verstegen deserves special thanks for his willingness to join this project at short notice and work to a nearly impossible deadline. Finally, I am grateful to Huguette and Yves Boulard for providing the space for me to write in Le Havre. Lily, this project has coincided with the first three years of your life and has no doubt been influenced in indefinable ways by your presence: I dedicate this book to you.

Richard Hoffmann would like to thank Jószef Laszlovszky for providing him with his first formal opportunity and obligation to think holistically about the medieval environment and also Verena Winiwarter for gently correcting some of his more egregious errors as he did so.

Wybren Verstegen wishes to thank Richard Hoffmann, Jan Willem Oosthoek, Mark Stoll, and Tamara Whited for their comments on the draft version of chapter 4.

PREFACE

Just behind the old neighborhood of Saint-Leu in the northern French city of Amiens, a capillary network of canals shapes a watery enclave within view of the mighty Notre-Dame of Amiens, the city's towering gothic cathedral. The canals and the fertile parcels bordering them, covering some 300 hectares today, seem all but subtracted from the surrounding cityscape, and they are in part older than the cathedral itself—even the first version of the edifice that was standing in the year 850 CE. Seemingly a world apart from the human activity of a prospering twenty-first-century city, the canals feed the Babylonian luxuriance of the "floating" gardens bordering them. The visitor may feel the brush of an overhanging willow branch against a cheek as the narrow boat, provided by a conservation association and equipped with a silent, nonpolluting electric motor, wends its way through a small part of the labyrinth open to tourists. A few paces away from rush-hour traffic, warblers, wrens, nightingales, woodpeckers, thrushes, kingfishers—in all, some sixty species of nesting birds—find habitat among the rushes and reeds, club moss, chickweed, and water lilies that thrive in and near the water.

Yet the canals and the parcels they enclose are the historic heart of Amiens. The zone is known as *Les Hortillonnages*, a name formed from *hortus*, the Latin word for "garden," and *aire*, Latin for "small surface." Indeed the ancient Romans occupying the northern frontier of their empire organized cultivation of the area in order to feed their troops, but human groups had likely occupied the site far earlier, subsisting from the varied resources of a marshland ecosystem formed by the Somme River, its tributary the Avre, and their many branches. Centuries after the Romans exited, medieval farmers fashioned Les Hortillonnages by draining and canalizing, thereby transforming over 300 hectares into cultivable plots delimited by the canals. As was true of other northern French cities, canals became the "economic nerve center of preindustrial urbanization" in Amiens, irrigating the vegetables that fed the city and providing water for the dyers who set up shop along their banks. (Guillerme 1988, 52) The people of Amiens also extracted the peat that had formed slowly from decaying plant life, drying it and burning it as fuel. Until the early modern era, the market gardens covered an expanse of over 1,000 hectares in all, enveloping the city that grew in their midst, depended on them, and eventually pushed them aside.

Les Hortillonnages could never be a permanent construct, solid like the cathedral. The soil is fragile due to its peaty and alluvial structure and the high water level. The banks are under continual threat of erosion and require assiduous upkeep. As the marsh tries to reassert itself, so too have human activities limited the gardens' surface area over the last half-millennium. A citadel built in the sixteenth century cut off part of Les Hortillonnages to the north; in 1825 the Somme underwent canalization, although a few branches of the former river were retained to feed the smaller canals; the coming of the railroad in the 1840s separated the community of Longueau ("Long Water") from Les Hortillonnages to the southeast; and in the twentieth century a new boulevard sectioned off their northwest portion as well as a marsh. Just as Les Hortillonnages were becoming an obstacle to urban development, local market gardening declined as efficient but energy-intensive agriculture and transportation lowered the cost of produce grown as far away as the South of France. Many of the vegetable plots gradually gave way to suburban idylls, complete with lawns, geraniums, and garden gnomes. Then in 1975, with a projected ring road about to deliver the *coup de grâce* to Les Hortillonnages, an association galvanized by Nisso Pélossof, a man of Greek origin, mobilized support to save the site. With a local membership of 1,300 or so, the body has since become the main actor both in conserving the canals' ever-crumbling banks and in sensitizing the public to the delicacy of this landscape crafted jointly by nature and human effort. In 1991 France's Ministry of the Environment classified Les Hortillonnages under the revealing label *Paysages de Reconquête* ("Reconquered Landscapes"). In the case of Les Hortillonnages, "reconquest" implies (in unintended fashion) the paradoxical preservation of a landscape from its two creators—water and humanity.

The story of Les Hortillonnages echoes the larger environmental history of northern Europe, with such features as the omnipresence of water, the ancientness of its human manipulation, ecological vitality permitted yet circumscribed, small-scale agriculture that produced three crops per year, threats from the transportation networks born of the Industrial Revolution, and a conservation group determined to maintain the integrity of a particular ecosystem. In opening this volume with such an example, I intend neither to play down the uniqueness of this local story nor to imply that one can find the same ecological components and historical developments replicated throughout the northern portions of the European subcontinent. Urban canals in themselves can be found in many towns and cities from Normandy to the Baltic Sea, but they obviously do not characterize the settlements of alpine Switzerland. The revolutions in agriculture and transportation also took place in Britain and Germany but less so in northern Scandinavia. Still, if taken as an analogy rather than as a microcosm, the story of Les Hortillonnages hints at wider patterns of reciprocal interaction between nature and culture over time—environmental history— that occurred in northern Europe.

What are the boundaries of this region? We are defining it essentially by latitude, covering the British Isles and the European landmass between approximately forty-five and sixty-five degrees north, and as far east as about the Oder River. Stopping at the Oder, which has provided much of Poland's western border since the end of World War II, does not do justice to ecological patterns, for the great European deciduous forest extends all the way to central Asia. Rather, our choice reflects in part the conventions of historical literature and is admittedly one of the limitations of this book. We do make occasional forays beyond the Oder, but readers should not look for sustained discussions of the eastern Baltic basin, much less areas further east. In terms of present-day nation-states, our region encompasses Ireland, Great Britain, the northern half of France, Belgium, the Netherlands, Luxembourg, Iceland, Norway, Sweden, Denmark, Finland, Germany, Switzerland, and the alpine portion of Austria. Given our geographical scope, our time frame covering all of human history, and the weighty historical literature on these "old" nation-states, this volume was best handled as a collaborative project. The combination of individuals' expertise represented here is, we believe, one of its strengths. Tamara Whited wrote the first two chapters and most of the last two chapters; Richard Hoffmann contributed chapter 3 on the Middle Ages; Wybren Verstegen wrote chapter 4 on the Early Modern period and the case study on fuel in the Netherlands; Jens Ivo Engels and Hilde Ibsen contributed to chapters 5 and 6, and each authored a case study on, respectively, the German Greens and Scandinavian trends in conservation and sustainable development. All authors contributed to the bibliographic essay.

As is true everywhere on earth, latitude and topography have primarily determined northern Europe's climate and biomes—geographic regions characterized by distinct biological communities—but in the case of northern Europe special mention must be made of bordering seas that separate Britain from the continent and cut deeply into Scandinavia. Much of this region therefore has an oceanic climate, with relatively even precipitation throughout the year and reasonably gentle seasonal changes. Regions further east feel the greater extremes of typically continental climates, although the oceanic influence stretches to central Europe. This is due to the east-west topographical arrangement of Europe, allowing cool Atlantic air unimpeded passage well inland. The largest geographic units of northern Europe are its lowlands, namely the North European

Plain, extending east-west on a continental scale. So, too, do the mid-altitude mountains—the German mountains south of the North European Plain, the Vosges, the Jura, the Sudetes—which lie for the most part along an east-west axis, and the great mountain wall of the Alps acts as the southernmost axis, creating a climatic division between northern and southern Europe. The Alps and their mid-altitude neighbors not only helped structure the climate and distribution of plants and animals in northern Europe but also conditioned the patterns of trade and communication slowly carved by Europeans since the Neolithic Era. Alpine passes and the Rhône River corridor allowed a direct north-south trade route in amber and metals from the Bronze Age. The links between northern and Mediterranean Europe form a profound historical continuity, most decisive for the eventual nation-state of France (as tomatoes from Provence in the markets of Amiens attest) and deserving of a larger, transregional environmental history. Northern Europe's rivers and their valleys likewise played a major role in shaping biomes and communication routes; some ten major rivers flow approximately northwest, emptying into the Bay of Biscay, the English Channel, the North Sea, and the Baltic Sea, and linking northern and southern subregions. This book will not treat the Danube River and its basin, for though the river's headwaters form in the Alps, its environmental history is more intimately linked to the Balkans, the Black Sea, and ultimately the Mediterranean.

With their inevitable variations and shadings, northern Europe contains two characteristic biomes. The larger of the two is a mixed deciduous forest, halted by cold winters at approximately sixty degrees north and in mountainous areas, thinning out in the more arid climate to the south. In the northernmost latitudes lies the boreal forest, with fewer, coniferous species, though many of its animal species also thrive in the deciduous belt, from large ungulates such as elk and deer to small furbearers. These biomes have changed greatly over time, however, and they can be fully described only by including histories of extinction (of aurochs, brown bears, and wolves, for example) and countless species introductions (from woolly sheep in the later Neolithic to Douglas fir in the last two centuries). Although the more spectacular climate changes have had an impact, human drives to use and transform nature have left profound marks upon many northern European ecosystems, ever since the hypothesized overkill of "megafauna"—woolly mammoths and rhinos, for example—in the final millennia of the Pleistocene. Our collective intention is to introduce readers to the various and evolving adaptations to northern European biomes, that is, to the northern European way of mixing human needs, desires, and skills with an evolving natural inheritance.

By way of a quick preview, two fundamental examples show the importance of an environmental perspective for fully understanding the key turning points in the region's history. First, the gradual and solid hold of agriculture, an export from the Middle East, occurred in the context of favorable climate, evoked above, and generally rich soil. Implanted in the Neolithic, agroecosystems became embedded by the late Bronze Age. This was advanced solar agriculture, characterized by dry farming, heavy reliance on grains, the use of animals for traction, and family-based farming, with some common access to resources, largely for subsistence but also for production for markets, which increased over time. Drainage was more of a concern than irrigation in most areas, and scrub and forest provided pastoral and other resources that helped complete the system. Northern European farmers in conjunction with bad weather could occasionally and unwittingly bring catastrophe upon themselves in the forms of dearth and sometimes famine, yet the traditional pattern of farming had a remarkable staying power over some two-and-a-half millennia.

More recently and in a more compressed time frame, northern Europeans long incubated and then hatched the Industrial Revolution. Based on fossil fuels and mechanization, modern industry brought about a host of oft-reviewed environmental changes. They ranged widely, from the emblematic slag heaps outside the mines to more intricate ecological shifts, as in the chemistry of rivers. The Industrial Revolution linked areas formerly unknown to one another, it fed the growth of cities, and it caused both human and animal depopulation of rural areas. Northern Europe proved to be a relatively small sink for its own industrial effluents, a basic factor in the emergence of a vibrant environmental movement in the later twentieth century. And of course, northern Europe's Industrial Revolution not only harnessed much of the earth to its demands for resources and markets, but it also provided a template and set of values aggressively emulated on every continent. Patterns of ecological degradation initially experienced in northern Europe became a common industrial legacy, one shared especially by peoples and places in North America and East Asia. Just as the Middle East's outstanding material legacy is agriculture, northern Europe's is surely industry, though one fraught with far more questions about its sustainability based on historical or current patterns.

With these two signposts established, we invite our readers to immerse themselves in the larger story of the northern European home and its successive remodelings.

September 2004

Le Havre

ADAPTATIONS ACROSS MILLENNIA
Northern Europeans of the Paleolithic and Mesolithic Eras

Over the last several centuries, Western culture has imagined its European homeland as supremely suitable for human habitation. Eighteenth-century concepts of a picturesque landscape were combined with descriptions of other continents, which were often rooted in imperialist conquest. This produced a portrait of Europe as seasonal without blistering extremes, fruitful, and boasting a mostly moderate topography. Europeans lived comfortably far from the climatic rigors of the tropics and the poles; European explorers would negatively connote even the riches of the Americas as "wild," whereas Europe was pleasingly domesticated. To place these assumptions under a critical lens, we begin by focusing on the Paleolithic Era, an archaeologist's term referring to the longest period of human history, which dated from 2.5 million years ago in Africa and about 900,000 years ago in Europe west of the Caucasus. A broad overview of these many millennia provides a foundation for understanding European adaptations to changing physical environments.

For if we push our story back to about one million years ago, Europe was anything but hospitable. Hominids had evolved in the tropical and subtropical environments of Africa, migrating into the Middle East, North Africa, and the Caucasus by 1.8 million years ago. This relatively rapid migration, which would eventually affect all land masses of the planet, was then stalled at the eastern edge of Europe for at least a half-million years. No other continent of the "Old World" posed as daunting a set of barriers to human settlement, up to a point, that is, in hominid evolution.

FROM THE FIRST EUROPEANS THROUGH THE LONG REIGN OF THE NEANDERTHALS

Various arguments have been advanced to explain the delay. The waters of Gibraltar and an enlarged Caspian/Black Sea may have hindered migration into Europe, yet these were permeable barriers, given the many climatic changes during the half-million years in question. Geologists and paleoanthropologists have also wondered about the possibility of an earlier European colonization; evidence of human presence has not been as well preserved in Europe as in Africa due to the scouring of glaciers and the tendency of European rivers to re-cut their channels, thus destroying ancient remains. A stronger argument points to the greater difficulty of colonizing Europe as opposed to Africa and large parts of Asia. Some three million years ago, there was an acceleration of a general cooling trend that had begun to affect planet earth much earlier. The Pleistocene geologic epoch dates from the onset of major glaciations, from about 2.3 million until 10,000 years ago. Forming at the poles and in alpine reaches, ice exacerbated the rigors of winter and seasonal contrasts that were al-ready Europe's lot due to its latitude, although glaciers did not cover Europe during the entire Pleistocene. Lower temperatures changed the flora and hence the food available to mammals. Relatively low annual mean temperatures and extreme seasonality posed distinct challenges; coupled with knowledge of the adaptability of primates, environmental conditions seem to explain the absence of hominids in Europe for over two-thirds of the Pleistocene. Humans of one million years ago were fully cultural beings, as ample archaeological evidence has shown, yet their culturally transmitted behaviors were barely up to the task of settling glacial Europe of the early Pleistocene.

An intriguing thesis advanced by A. Turner posits that early humans were kept out of Europe by the density of large carnivores and carcass destroyers, namely giant hyenas and saber-toothed tigers (Roebroeks and van Kolfschoten 1995, 288). These animals posed a direct threat to human ancestors and were better hunters and scavengers of herbivores. Between 600,000 and 400,000 years ago, these animals became extinct in Europe, leaving only smaller populations of leopards, lions, spotted hyenas, and wolves. Once the most dangerous carni-vores had left the scene, the possibilities opened up for hominids to hunt and scavenge. Over time, the diversity of herbivore populations on the European tundra and steppe grew to include rhinoceros, large bovids, horses, and other species, yet the very richness of animal resources, and the relative poverty of plant resources, was itself a challenge: greater organizational skill is required to hunt large animals than to gather plants. Finally, the very scarcity of hominids, even in Africa, helps to explain a gradual colonization. In the case of Europe, scholars have hypothesized the isolation and demise of small, founding groups, and thus the recolonization of the continent several times. In the absence of specialized skills, humans were likely to enter and exit northern latitudes periodically.

The earliest human remains recovered in western Europe are those at Gran Dolina in Spain, dating to 900,000 years ago, but most undisputed dates for early human occupation of northern Europe are less than 400,000 years old. One of the earliest sites in our region is located at Bilzingsleben in the Wipper Valley of eastern Germany. Some 350,000 to 400,000 years old, the human remains are those of either *Homo erectus* or *Homo heidelbergensis*. Evidence found at Bilzingsleben includes organized campsites; bone, wood, and many stone artifacts; and evidence for the specialized hunting of large animals, namely rhino. Significantly, animal bones reveal that bear was the only carnivore present.

The Bilzingsleben II site is associated with an interglacial epoch, a period of warmer climate in between glacial phases. Much of the west-central European corridor, the area between the Alps and the North Sea, was in fact never subject to glaciation, a circumstance that allowed early occupation as well as the ultimate preservation of sites. Whereas southern Europe has yielded human remains dated to the fully glacial climate of 200,000 to 300,000 years ago, occupation of northernmost Europe remained intermittent until the final retreat of ice at the end of the Paleolithic. For example, *Homo heidelbergensis* settled in the British Isles but may have abandoned them, along with northwest continental Europe, during the last glacial maximum, the final period of intense cold lasting from 20,000 to 16,000 years ago. John Campbell has rejected this "hiatus hypothesis" for Britain, preferring to see early hunter-gatherers as able to adapt to a colder climate within a few thousand years. Provocatively, he writes that modern assumptions about human comfort levels have warped our understanding of hunter-gatherers. As long as there were opportunities to hunt and fish at the edge of the ice, periglacial Britain may have remained an acceptable environment to humans during the last and coldest phase of the Paleolithic. (Roe 1986, 810) Even if life had been marginally possible there, the environs of the glaciers hardly beckoned migrants, so what kinds of environments initially attracted humans to northern Europe? The answer is laced with conjecture, yet techniques such as pollen analysis can help reconstruct the kinds of environments that early Europeans favored. Several specialists on the European Paleolithic have hypothesized an early human preference for mosaic environments—that is, areas of mixed rather than uniform vegetation. Mosaic environments offer at least three advantages: first, as ecosystems they are particularly resistant to various disturbances, such as changing climate and disease; second, local disturbances would require only short migrations on the part of humans and other animals; and third, diverse flora provide ecological niches for a diversity of fauna, namely grazers. Such was Europe north of the Alps and the Carpathians, during both colder and warmer, more forested phases. Along with the west-central corridor, the western end of Europe was particularly favored: the Atlantic Ocean's moderating effect on climate reduced seasonal extremes and thereby allowed a more concentrated, diverse pattern of vegetation. Rich flora invited animal populations; with faunal migrations from the east occurring from the Middle Pleistocene, forest and steppe mosaics became even more productive. Through proximity to southern European refugia—areas where species are able to survive despite extinction in nearby regions—western Europe also benefited from the rapid recolonization of species during interglacials.

Within the mosaics, the earliest occupiers of Europe chose to live both at "open" locales along ecologically rich streams, lakes, and marshes and at "closed" locales, such as caves and rock shelters. Humans of the Lower Paleolithic lived in very small groups at these sites. They have left no evidence of complex material structures, and samples from several dozen sites across Europe show that they used local raw materials only, obtaining them usually within a radius of thirty kilometers of a campsite. The full range of their activities at these sites remains obscure, yet Bilzingsleben and other sites suggest that they were butchering and processing animal carcasses as well as selecting and knapping stone. Activities implying more integrated social groupings, such as transmitting skills and sharing food, can be imagined if not absolutely inferred from the evidence.

In the millennia between 300,000 and 250,000 years ago, a European descendant of *Homo heidelbergensis* appeared: *Homo neanderthalensis* occupied the continent until modern humans began to replace them 40,000 years ago. The archaeological record begins to reveal changes in behavior that in turn denote an altered relationship between early Europeans and their environments. Neanderthals adapted to and even flourished in periglacial surroundings, including a northern Europe often devoid of tree cover but abounding in permafrost. With the advent of Neanderthals, long-term occupation of sites becomes apparent. If unable to adapt to all European environments, Neanderthals did, broadly speaking, come to terms with cold and altitude.

One set of adaptations was genetic. Neanderthals were thick-boned, extremely muscular people, but otherwise with much the same anatomy as modern humans. Their skulls had long faces, large brow ridges, and huge noses. Their bones also show evidence of frequent fractures and degeneration, and aging techniques indicate that the oldest Neanderthals rarely lived beyond their

Illustration of Neanderthal encampment. (Time Life Pictures/Getty Images)

early forties. Hunting large animals such as elephant and rhinoceros was a dangerous business; the apparent frequency of physical trauma leads one specialist to liken the lives of Neanderthals to those of "modern rodeo riders." (Klein 1999, 475)

The term "Mousterian" describes the material culture left by Neanderthals starting from 250,000 years ago. The essential industry of these Middle Paleolithic people was flake technology, the fashioning of tools from "flakes" chipped out of stone cores such as flint. The Mousterian was a period of little or no innovation across a span of 200,000 years; there are few artifacts of bone, antler, or ivory and no artistic production. Overall, Neanderthals kept to a small variety of artifacts and usually derived them from raw materials transported a few kilometers. Through one kind of lens they appear conservative and lacking in innovation, especially when compared with their modern successors; through another, they show a successful adaptation to changing environments.

Neanderthals used flint choppers to process carcasses, and they probably used tools such as wooden spears to kill animals, in particular reindeer, red deer, horse, aurochs, and steppe bison. The extent to which Neanderthals scavenged opportunistically or deliberately hunted is difficult to infer from the remains at archaeological sites. Some of the richest sites occur in the valley of the Dordogne River and its tributaries in southwestern France; though off-center as far as our concern with northern Europe, the finds from the Dordogne may have much to say about the ecological niches of Neanderthals elsewhere. Specialists concur that Neanderthals hunted reindeer, given the presence of large, meat-bearing reindeer bones at the Dordogne sites; carnivores would have had first pick at these parts had the Neanderthals only scavenged. Evidence for hunting of large bovids, the aurochs and bison, is less conclusive but suggestive: at four sites in France, remains of aurochs and bison account for 93 to 100 percent of all animal bones present, showing a degree of specialization that is difficult to reconcile with scavenging. A strong case has also been made for the hunting of bovids, horse, and reindeer by means of driving herds over cliffs. The larger animals were butchered close to if not at the kill sites, and the remains of skulls and jaws point to brains and tongues as coveted sources of nutrients.

Evidence points strongly toward the eventual replacement of Neanderthals by modern humans. The earliest modern humans evolved in Africa and appeared in Europe about 40,000 years ago. The astonishing recovery in 1997 of mitochondrial DNA from a Neanderthal's humerus shows that the lines leading to each subspecies split very long ago—on the order of 550,000 to 690,000 years before our time. Once the moderns appeared, hybridization might have been possible but was unlikely. Such a question cannot be tested in the absence of DNA recovery from fossil remains of the earliest modern humans. At any rate, different species within the genus *Homo* adapted to the cold Europe of the last glacial era, overlapping across time and, to a lesser extent, across space.

Neanderthal populations may well have begun to contract before the arrival of *Homo sapiens*, who entered Europe from the Middle East and gradually colonized the continent between 40,000 and 28,000 years ago to the detriment, and ultimate extinction, of Neanderthals. Regional coexistence is extremely difficult to pinpoint; given the limitations of radiocarbon dating, two groups in the same area who appear to overlap may have actually missed each other by a century or so, or vice versa. Small areas of coexistence have been identified with reasonable surety in the southern half of France and southwestern Spain, and evidence for acculturation may be detected in the technologies corresponding to the temporal overlap. To a modest extent, then, one can see a partially shared adaptation occurring in pockets of western Europe, with Neanderthals picking up techniques from the moderns.

Paul Mellars, a leading specialist on the Neanderthals, believes that modern replacement was extremely gradual and therefore unlikely to have occurred through violent confrontation. But why did modern humans and not Neanderthals survive? The answer to this question is not as obvious as it may first appear. Physically, the Neanderthals had bodies that were better adapted to the periglacial environments of Europe, yet a few adaptations in modern humans may have made all the difference, giving them a markedly greater capacity for culture. They did not migrate out of the Middle East until the occurrence of a crucial set of technological and cultural changes—the "Upper Paleolithic Revolution."

MODERN HUMANS IN EUROPE: NEW TOOL KITS AND SURVIVAL STRATEGIES

The modern colonization of Europe took place during the latter part of what paleoclimatologists refer to as OIS 3, an oxygen isotope-based calendrical marker for the period between 59,000 and 24,000 years ago. Overall, the ice stabilized during OIS 3, covering the Scandinavian mountains but stopping short of the west coasts of Norway and Denmark, most of lowland Sweden, and southern Finland. Large herbivores thrived on the steppe and tundra, presumably easing life for the Neanderthals. The warmest part of this period began about 43,000 years ago, but a few thousand years later—just as modern humans were making their European entry—average annual temperatures in Europe began to oscillate with very high frequency. Temperatures rose five to six degrees Celsius every thousand years or less, then plunged downward again, as evidence from Greenland ice cores has confirmed. The following 15,000 years were more mild on average but also extremely unstable. Temperate woodland crept into steppe and tundra, facilitating the expansion of the migrants, who advanced into Europe along the valley of the Danube River, or perhaps along the Mediterranean coastline. The temperature oscillations may have fatally destabilized Neanderthal populations, known for their generally conservative behavior. Modern humans arrived in Europe precisely during an era which placed a premium on the ability to change subsistence strategies and technologies in order to adapt to changing environments.

The first modern Europeans remained, of course, Paleolithic peoples, living by hunting and gathering. On the one hand, many ethnographic studies have concluded that gathering contributes more to diet than hunting. Furthermore, hunting is surely overrepresented in the fossil record, for only plants that are roasted or parched can become evidence for the archaeologist; plant sources ranging from leaves and fruit to roots and tubers rarely leave traces. On the other hand, game animals are more abundant at middle and upper latitudes. In glacial periods, the European steppe and tundra were productive enough to support voracious, large animals such as mammoth, but flora was not diverse enough for human needs. Europeans of the Upper Paleolithic may well have subsisted almost entirely on meat. Evidence from Germany and southwestern France points to an overall continuity of prey species from the Middle to the Upper Paleolithic, with red deer, reindeer, and horse dominating the record. In those areas of possible overlap, modern humans clearly specialized in reindeer hunting, whereas the Neanderthals remained generalists. In later, colder times, however, the abundance of reindeer bones may simply indicate the greater survival of the species rather than the specializing of hunters.

Overall hunting strategies appear to have changed the most, followed by technology. As had the Neanderthals before them, modern humans hunted large animals with great success. Though highly challenging, a kill would bring huge amounts of meat to typically small groups; as needs and technology developed, those same animals provided hides, sinew, bone, antler, ivory, and dung for a multitude of other objects and building purposes. Small, mobile subsistence groups, or perhaps even smaller procurement parties, most likely followed migrating herds of herbivores, although to a debatable extent. Reindeer can cover up to thirty kilometers in a day over rough terrain, and herbivores alter their migration routes from one year to the next. Only specialized hunting parties likely intercepted herds, a skill requiring not only advance planning but also a sophisticated ability to mentally map landscapes, animals, and neighboring groups of people. Mapping and monitoring the environment may have been one of the crucial skills, allowed by neurological changes, which gave modern humans an adaptive edge in glacial Europe.

Technology allows the greatest insight into the subsistence patterns of the newcomers. The Upper Paleolithic emerges as the "quantum change from everything that went before." (Klein 1999, 522) Industries of the Upper Paleolithic changed within the same environments and climatic phases; conversely, the Mousterian remained continuous across changing environments and climates. Such facts caution against establishing simple, deterministic relationships between technology and the environment, suggesting a portrait of modern humans as unrelentingly innovative. They manufactured many more kinds of tools, tools used to make other tools, and composite tools whose components were held together with glues and thongs. They chipped stone tools from blades—tools at least twice as long as they are wide, with much more cutting edge than a flake. (Neanderthals had also made blades, though fewer of them and in a different way.) In a now more complex process of production involving more subtle hand movements, the desired tool would emerge from the blade after retouching—chipping with stone, antler, or wooden hammer. Stone tools included scrapers, perforators, and burins, tools with chisel-like points. Carved and polished from bone, antler, and ivory, nonutilitarian items also characterize the Upper Paleolithic.

Modern humans evidently used and perfected their toolmaking skills in part to become better hunters. One of the key innovations was the replaceable spear tip made of bone or antler. With a split or beveled base to attach to the spear, points abounded in enormous variety across space and time. They made for cleaner, surer, less dangerous kills, in obvious contrast to the Neanderthals' short spears with broad points. Moderns used backward-facing barbs to spear fish or animals crossing streams, while snares and traps reveal the incorporation of smaller game into the European diet. By 20,000 years ago, Europeans were fashioning sophisticated spear throwers, with a handle at one end and a hook at the other to attach to the spear; this tool greatly increased the hunter's throwing range, facilitating the kill and keeping the hunter more or less hidden, and protected, from large animals. Conclusive evidence for the use of bows and arrows dates only from the very end of the Paleolithic, some 12,000 to 10,000 years ago, in northern Germany and France.

Specialists note that across the cultural periods of the Upper Paleolithic, much variation among artifacts is stylistic in nature; this clues us in to a very modern aspect of culture, that is, fashion. Change for its own sake may indicate a growing flexibility with respect to a challenging environment. In this vein, the production of art from the outset of the Upper Paleolithic shows a striking turn of human resources away from subsistence. Portable art, whether carved or engraved from bone, ivory, or animal teeth, has been recovered from westernmost Europe to Siberia, and paintings remain in some 200 caves in southwestern Europe. The Magdalenians achieved the well-known paintings at Lascaux, Niaux, Altamira, El Castillo, and other sites, but paintings at other French caves, namely the spectacular Chauvet Cave near the Rhône River, place the debut of art beyond 30,000 years ago. The purpose of the paintings cannot be fully inferred, and it is not known whether artists worked in hopes of assuring successful hunts (most images on cave walls are of animals), marking territorial boundaries, or recording social structures, worldviews, or spiritual visions. Perhaps all of those purposes, in addition to others, drove the production of art. We might follow Randall White's assertion that these Europeans were beginning to produce "their own world, a symbolic world that included imaginary creatures and deities." In his view, even the most naturalistic animal images convey "a set of ideas and conceptions that are purely cultural." (White 1986, 104)

Two additional technological achievements further suggest the glimmerings of control over nature: lamps, most often made from a slab of limestone, with animal fat for fuel and moss for wicks, and cobblestone fireplaces, with wood as the usual fuel. Europeans regularly constructed hearths from 20,000

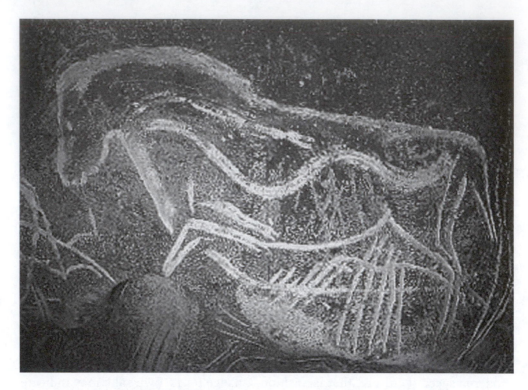

Painting of a horse found at the Chauvet Cave in France, discovered in 1994.
(AFP/Getty Images)

years ago, but their very distant ancestors had already captured and used fire. It is extremely difficult to differentiate between natural and anthropogenic fire when studying the Paleolithic and Mesolithic eras; however, an excavation in Brittany has yielded a hearth dated to approximately 465,000 years ago, and one at Torralba, Spain, shows evidence of fire drives for purposes of hunting some 400,000 years ago. In other words, early hominids in Europe manipulated fire. *Homo sapiens,* too, engaged in fire-assisted hunting, benefiting from both the fire drive and the subsequent return of fauna to young growth.

The use of fire influenced where people lived. Its mastery facilitated cave dwelling by protecting human occupants from the intrusions of carnivores; caves, reciprocally, would have helped to keep fuel dry, even though smoke prevented fire-building beyond the mouth of the cave. Caves also provided protection from rain, snow, and wind, and they retained a nearly constant temperature year-round. Upper Paleolithic Europeans occupied cave mouths and rock shelters where available, but such was not the case over most of their range. In central and eastern Europe, large depressions with postholes, stone blocks, and large bones suggest built dwellings, as do the ancient fireplaces found at these sites. Large open-air settlements, usually located near water and composed of multiple units with hearths, were a novel form of settlement by about 28,000 years ago. Whether or not human groups occupied the large sites all year gets back to the question of whether they were able to follow migrating herds. Slight evidence of smoking and freezing meat occurs at open-air sites in eastern Europe, implying year-round residence as well as delayed consumption of food. Finally, the existence of both large and small settlement sites within the same region may indicate flexible social organization based on the seasonal availability of food.

The subsistence strategies, technologies, and social organization of modern humans were all put to the test during the final phase of the Pleistocene, a period from roughly 22,000 to 16,000 years ago dubbed the "last glacial maximum." Marked by extreme cold and aridity, the European climate made significant portions of the continent uninhabitable. The extent of ice speaks for itself: a mile-high sheet of ice covered most of northern Europe, including three-quarters of Ireland, two-thirds of Britain, all of Scandinavia and Jutland, the northeast of Germany, the northern half of today's North Sea, and all of the Baltic. In addition, glaciers formed in the mountains at more southern latitudes—the Alps, Pyrenees, Massif Central, and the mountains of Spain and Italy. The Scandinavian and Alpine ice sheets came to within 600 kilometers of each other; in between lay eastern France and central and southern Germany, some of the coldest parts of Europe, an arctic desert that was ice-free but nearly without vegetation. So much water was locked up in ice that ocean levels dropped 100 meters on average: dry land stretched from today's Bering Sea to the English Channel.

As the ice advanced, the forest evaporated, leaving only pockets of tree cover within the steppe landscape of central and even much of southern Europe. Floral diversity diminished, thereby reducing the diversity of medium-sized prey such as aurochs, elk, roe deer, and pig. Reindeer and horse remained throughout much of unglaciated Europe, yet these species were highly sensitive to short-term fluctuations in the weather. Human survival became impossible on the northern fringes of Europe and tenuous in much of the center of the continent. Lower temperatures year-round, winters that became far more severe, and scant resources forced a movement of humans and other species into southwestern Europe, a refuge area. For 10,000 years in many regions, parts of northern and central Europe became a no-man's land.

In light of climatic deterioration and consequent migration, we might expect to see only demographic contraction and cultural stasis. Instead we look to the late Upper Paleolithic for glimpses of new and lasting adaptations to the cold. Changes in hunting technology and strategy, clothing, and possibly transport occurred not only in the southerly refuge areas but also in extreme eastern Europe,

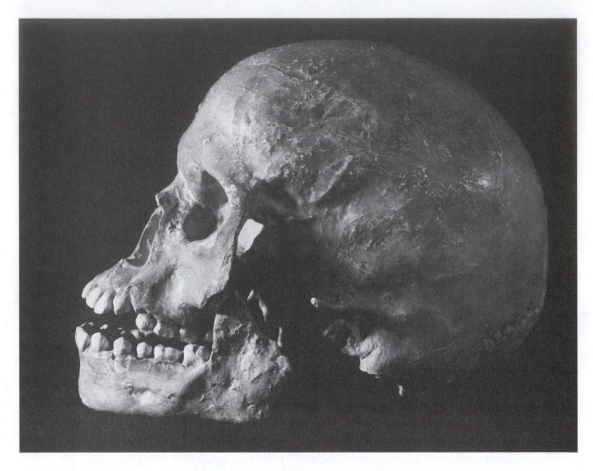

An Upper Paleolithic skull found at the Grimaldi Caves in Italy. The age of the skull remains uncertain, but it is probably Gravettian (30,000 to 22,000 years ago). (Archivo Iconografico, S.A./Corbis)

whose first human occupants arrived during these inclement times. It is there, for example, that we see the first evidence of regular fur trapping, namely of wolves and arctic foxes. Modern humans in eastern Europe were using bone needles with eyes and fine thread by about 20,000 years ago; burials at Sunghir, located northeast of Moscow, show partial fur suits that were not only tailored but even decorated with ivory beads. The fully tailored fur suit probably first clad a European somewhere between Sunghir and Germany. Figurines from a number of eastern European sites are themselves well clad. Thus, by the last glacial maximum, humans managed to protect their bodies from the cold; consequently, caloric needs did not necessarily rise much in the colder weather. Where human groups were able to subsist, they did not cease to live in large, open-air settlements during the icy end of the Paleolithic. Evidence for new modes of transportation remains speculative, but Europeans possibly developed sledges, skis, and showshoes at this time. At any rate, they did have the fibers, grease, and lashed frames to fashion such new means of moving over ice and snow.

NORTHERN EUROPE AFTER THE ICE: HUMANS ENCOUNTER A NEW ENVIRONMENT

Ice began its slow recession in northern Europe 16,000 years ago and was in full retreat by 10,000 years ago. Stephen Pyne characterizes the Europe that emerged with a startling comparison: " . . . the retreating ice made Europe a virtual *terra nova*. Old World Europe was, paradoxically, as much a new world as the Americas, and certainly newer than Australia and Africa." (Pyne 1997, 18) Northern Europe as it emerged from the last ice age is, in terms of topography, climate, soils, vegetation, and fauna, essentially the Europe of today. Melting ice altered the very shape of Europe: the rising North Sea severed Britain from the European continent by 7,800 years ago. The waters also caused rivers to back up, peat to accumulate, and low-lying plains to become wetlands, as in eastern Britain and the low countries. Released from the weight of ice, the British and Scandinavian land masses rebounded, thus slightly counteracting rising sea levels. As had earlier glaciations, the last one carved the major mountain ranges, namely the Alps, with U-shaped valleys, and the glaciation also left fjords—narrow, deep inlets—that met the sea along the coasts of Norway, Denmark, and western Sweden. Retreating glaciers sculpted moraines and numerous lakes in Scotland, Sweden, Finland, northeastern Germany, and northern Poland.

One geographical particularity of Europe lies in the high proportions of water to land and coastline to land mass. The proximity of the seas—the Atlantic Ocean, Irish Sea, North Sea, and Baltic Sea—moderates the seasonal tendencies of northern Europe with the added benefit that the Atlantic brings the Gulf Stream's warm Caribbean waters close to Europe's shores. Holocene temperatures reached a maximum around 7,000 years ago, then stabilized over the next thousand years to create the climate Europeans have since known (excepting fluctuations such as the Medieval Warm Period and the Little Ice Age). Sea-level temperatures rarely exceed 20°C in summer or dip below freezing in winter. Northern Europe's oceanic climate distributes precipitation fairly evenly throughout the year, and due to the east-west arrangement of mountain ranges, oceanic effects are felt far into central Europe. While most of the Mediterranean risks seasonal water shortage, Atlantic Europe has to cope with too much: waterlogged soils, leaching of soluble nutrients below the reach of grasses, seasonal flooding, erosion, and deposition come with an overabundance of water.

Photograph taken from space of the North Sea meeting the coasts of Brittany, the Netherlands, and England with islands of the Rhine River in the foreground. (Time Life Pictures/Getty Images)

Large, relatively low-gradient rivers run all year, permitting navigation far into the interior and pouring nutrient-rich water onto wide continental shelves, where once-abundant aquatic life supported large-scale commercial fishing.

Neither the Atlantic nor the Mediterranean moderate conditions in continental Europe. Almost everywhere north of the Alpine-Balkan mountain ranges and east of the Bohemian upland, climates are extreme, regularly rising above 25°C in summer months and averaging below freezing in the winter. The upper quarter of the Scandinavian Peninsula lies above the Arctic Circle. Finally, the Alps have a climate all their own, for the mountains form a convergence zone among Atlantic, Mediterranean, and continental weather patterns.

The glaciers' greatest gift to all life forms, including humans, was arguably soil, though generations of deciduous trees would improve upon the glacial inheritance by contributing organic matter. Ice flowing over rock had created enormous dust clouds that were deposited as loess—friable, silty soil—in a swath of north-central Europe stretching from the Ukraine to the Paris basin. These several meters of topsoil would be used by a later wave of human migrants (see chapter 2). Adjacent to the loess belt is the North European Plain, an area of mostly thinner soils but also containing the rich alluvial floodplains of the Elbe, Oder, and Vistula rivers. Sturdy, rich soil did not underlie all of northern Europe, however; receding glacial meltwaters left acidic sands as they carved broad valleys in the northernmost part of the North European Plain, and acidic soils leached of their minerals (podsols) tended to form in areas under coniferous forest or heath, as in the northern, wetter parts of Britain.

Though unstable for many thousands of years, the newly exposed soils beckoned plant life from the south. Preceding the human migration out of the refugia, tree species advanced in waves, beginning with birch, aspen, and pine, then followed first by elm and hazel and later by oak, linden, alder, and ash. By 8,000 years ago, the patterns of woodland and forest that mark Europe today were in place. Much of central Europe became diverse, mostly deciduous, forest. Britain, which gradually became less ecologically diverse after its insulation, was dominated by oak forest in the lowlands and pine-birch woodland in the highlands. Scandinavia became the home of boreal forest hosting conifers such as spruce, pine, larch, and fir.

Animal populations followed the improved vegetation. Reindeer and wild horse sought out tundra as it receded to the north, their numbers giving way to populations of deer and wild pig as woodland ecosystems matured. Some species, such as elk and moose, declined in Britain's oak forests but thrived in the boreal forest of Sweden. Birds populated deciduous woodlands and established patterns of yearly migration. Whales and seals appeared off the west coast of Norway and in the Baltic. The chief incentive for humans to leave the southern refugia was to follow the herds of elk, reindeer, horse, and possibly mammoth, though the latter became rare in Europe by late glacial times. Humans recolonized northern Europe during the Mesolithic, the archaeological era succeeding the Paleolithic; the expansion began in the late glacial warming of 15,000 years ago and accelerated 10,000 years ago.

Mesolithic peoples remained hunter-gatherers but became more reliant on solitary forest species such as deer and wild pig. Hunting and butchering animals still required less energy than gathering and processing plants as a ratio to the amount of energy obtained from food. On the other hand, the far greater biomass available from edible plants in these warmer millennia was surely put to use. Both plants and animals dwindled in the lean season of late winter/early spring, a time when marine resources filled part of the gap for coastal inhabitants of northern Europe. In southern Norway, for example, where the

Mesolithic began around 10,000 years ago with the arrival of migrants from the North Sea as well as overland from Sweden, subsistence revolved around whales, seals, seabirds, and fish in addition to the usual red deer, wild boar, and moose. Depending on their location, Mesolithic peoples subsisted by foraging at several sites throughout the year, aggregating then dispersing according to the season, or remaining relatively sedentary.

Mesolithic Europeans altered their landscapes through fire more thoroughly than had their predecessors, thereby creating more predictable environments for themselves. Burning grasses helped to rejuvenate their environments over five- to six-year periods, attracting game, especially if open areas were maintained near water sources. Ian Simmons, an expert on the later Mesolithic, views the creation of large open areas as the most significant environmental legacy of the British Mesolithic; the full occupation of the mixed deciduous forest led to its manipulation, or "management." Europeans had learned to create the mosaics so favored by their Paleolithic ancestors. Manipulation could be taken to extremes: it was Mesolithic hunter-gatherers, not subsequent farmers and herders, who deforested the Western Isles of Scotland, which became treeless by 2,500 years ago.

Another crucial aspect of environmental manipulation concerns the fate in Europe of large mammals, or "megafauna." Did early Europeans hunt giant deer, woolly rhino, and woolly mammoth to extinction? Those three species disappeared entirely throughout their range; in all, Europe lost thirteen genera of megafauna at the end of the Pleistocene, with nine out of the thirteen surviving in Asia or Africa. While the extinctions were not as dramatic as those that took place in the Americas or Australia, scholars such as Paul Martin do attribute them to overkill by human hunters. Few extinctions of the late Pleistocene preceded human arrival, and extinctions occurred without replacement by other fauna. Martin also finds a fit between the relatively gradual pattern of extinction in Europe and the gradual spread of Paleolithic and Mesolithic hunters across the continent. Conversely, fairly sudden extinctions in the Americas followed the sudden appearance of hunters in the New World. (Martin and Klein 1984, 358–359; 388) Another explanation is basically climatic: cold-weather animals adapted to steppe and tundra could not survive in the woodlands that encroached in the wake of retreating ice. The early Holocene, with its more rapid climate change and more intense seasonality compared with previous interglacials, hampered adaptation by megafauna. However, certain warm-weather species, such as hippopotamus, should have returned as the European climate grew mild, yet they did not. Rejecting a possible synthesis between the two explanations, Martin notes that if climate and hunting had together threatened the survival of megafauna, then the extinctions in Europe

Illustration of two woolly mammoths walking through a snowy tundra. (Bettmann/Corbis)

should have affected more species and occurred more rapidly than they did. (Martin 1984, 289) The debate continues.

By the end of the Mesolithic, European hunter-gatherers had grown highly dependent on red deer, roe deer, and wild pig, maintaining grassland and forest edge, and perhaps scattering fodder, to attract these species. If living close to the seashore, they harvested shellfish and seaweed, and everywhere they processed any number of tubers, nuts, and other sources of plant food in ways that remain obscure. Just as opaque are the rituals and myths by which they understood and made use of these resources. These and other behaviors add up to a broadly based subsistence that took advantage of seasonal riches and relied on migration. Paleolithic and Mesolithic legacies bespeak significant change, but since they span nearly one million years, they also point to great continuity. Paleolithic Europeans occupied, retreated from, and reoccupied northern Europe depending on the extent of ice, adapting to and altering their environments within a given genetic and technological potential. At the risk of anachronism, the "sustainability" of these adaptations appears self-evident, though they were threatened by changes in climate or, in the case of the Neanderthals, the slow advance of modern human competitors. The succession of Paleolithic cultures created enduring ways of life that lasted across the millennia of glacial Europe.

2

EARLY AGRARIAN CIVILIZATION
From the European Neolithic through the Roman Era

After about 5500 BCE in north-central Europe, and 4000 BCE in Scandinavia and the British Isles, new elements begin to appear in pollen and archaeological records. Strictly speaking, the Neolithic Era (like the Paleolithic Era) is an archaeological term denoting assemblages of pottery and polished stone in addition to signs of permanent human settlement. Farming, indicated by the presence of cereal pollens and bones of domesticated animals, has often been assumed to be part of the Neolithic package. Direct evidence of agriculture has not, however, always been present at "Neolithic" sites, and therein lies a key problem: What does it mean if the package arrived in bits and pieces over long spans of time, as opposed to the steamrolling transformation suggested by the very name "Neolithic Revolution"? What are the implications of a possibly piecemeal shift for European societies and environments? At what point in the development of the Neolithic should we infer new social structures and values, new cosmogonies, and altered attitudes toward the natural world?

These questions are best explored through a regional tour of northern Europe, for the story is not always identical within segments of this large continent. A summary of the two most important debates concerning early agriculture will provide a backdrop for the tour. First, the question of causation has attracted scholarly attention since at least the early twentieth century: today the debate revolves not around the absolute origin of agriculture (which can be seen as a biological event) but rather around its diffusion throughout the world, implying far-reaching commitments by entire societies. Second, research on the mechanisms by which agriculture spread, whether through colonization or acculturation, has led to vigorous debate and perhaps a new consensus within European studies. Case studies of northern European regions will then tease out the implications of all of this for environmental history, and the chapter will conclude with a broad examination of the Bronze and Iron ages through the collapse of the Roman Empire.

WHY EUROPEANS ACCEPTED
THE RISKS OF AGRICULTURE

Nineteenth-century views of linear human progress held that the superiority of farming over hunting and gathering was self-evident and that, ecosystems providing, human groups would inevitably adopt agriculture once its advantages became clear. Charles Darwin held to this view, emphasizing that only the know-how of growing plants was the decisive factor in the spread of farming. Farmers produce more food than hunter-gatherers per amounts of space and time invested. Scholars rarely dispute this claim, but they have attacked notions of agriculture's inherent superiority based on discoveries of less robust skeletons and more disease, more social constraint, and higher work requirements among fellow travelers of the Neolithic and subsequent cultural eras. Many scholars find the assumed passivity of hunter-gatherers distasteful and unproven within both the linear-progress model and a second model that leans toward environmental determinism. Because the latter model continues to foster discussion, it deserves explanation.

A steadily deteriorating climate, dwindling plant and animal populations relied upon by Mesolithic peoples, or a population growing out of proportion to resources—such are the elements of more deterministic models hinging on external factors. In short, hunter-gatherers had little choice but to adopt a more productive mode of subsistence when the imbalance became too keen. In Mark Cohen's well-known theory (1977), population growth past the margin of subsistence is an ever-present danger for hunter-gatherers. Disaster was on the horizon, if not imminent, for Mesolithic peoples, and they readily adopted farming once the means became available. However, little concrete evidence for absolute resource shortage or declining health due to population pressure has been located either in the Middle East, where agriculture originated, or in Europe. Population levels remained low throughout Europe in the late Mesolithic and into the early Neolithic; most of the continent, away from coasts, large lakes, and rivers, remained uninhabited. In addition, the forests and seas of northern Europe provided generously in postglacial times. Indeed, agriculture emerged in areas where resources were ample and secure, whether one looks at the Middle East or Europe.

Likewise, little evidence for deteriorating climate has surfaced. Climate change is a slippery variable; it often appears to correlate with cultural shifts, yet causation is notoriously difficult to prove. A dramatic ecological phenomenon—the precipitous decline of elm trees in northwest Europe at about 4000 BCE—seems at first to support the case for a distinct climatic role in the diffusion of agriculture in Europe. Pollen data indicate that the elm population may have been reduced by half in some areas within a few decades. The transition from the "Atlantic" to the "Sub-Boreal" climate phases brought colder weather during this era; the change is, in some areas, synchronous with the beginnings of European agriculture. On the one hand, the same cooling that affected elm stands might also have strained the subsistence strategies of Mesolithic groups. On the other hand, some cold-sensitive tree species, such as ash, showed no decline, nor was elm decimated everywhere in northwest Europe, as one would expect to happen as a result of climate change. Furthermore, within most of the range of the elm decline, evidence for farming appears comfortably prior to 4000 BCE. Specialists now point more often to disease as the chief cause of the elm decline, though anthropogenic causes may also have played a role. In any case, climate seems not to have threatened the survival of Mesolithic peoples; conversely, the European climate may have become more conducive to farming. Central Europe of the sixth millennium BCE (the earliest traces of agriculture date from this era) was relatively warm and dry, much like the climate of southeast Europe and Anatolia. Agricultural techniques spread to central Europe from these regions.

Theories seeking to explain the diffusion of agriculture look increasingly away from environmental explanations and toward social and ideological factors. Domestication may have originated and spread among the most prosperous and technologically developed hunter-gatherers—groups already leading a semisedentary existence, trading regionally in exotic products and producing social inequality. The latter trend sparked the need for greater food production, as authorities sought to retain their prestige through periodic feasting and the possession of rare items. Trade likely accelerated the production of ceramics and stone tools. Whereas the evidence for feasting, trade, and inequality remains far from uniform across northern Europe, the appearance of houses of different sizes and megalithic graves possibly destined for select individuals do hint at social change during the early Neolithic. Unfortunately, the evidence does not tell us how to explain the rise of social inequality in the first place.

According to other theories, the widespread transition to agriculture had ideological causes. Biological domestication may have issued from new ideas and values centered on the domestic hearth, or *domus*. Houses and households preceded agriculture and on a symbolic level were its necessary precursors. Such an interpretation also suggests the prior existence of social inequality, for some members of the household were in greater positions of power to "tame" other members. In this view, agriculture involves a larger transformation of the wild into the domesticated. Once again, it is extremely difficult to trace the origins of such new cultural values in the archaeological record. Another perspective holds that households functioned less as symbolic units than as adaptive agents. Acting individually yet linked together through kinship bonds, house-holds produced a sort of kinetic energy as they progressed through their devel-opmental cycles, a sum of decisions that resulted in the rapid diffusion of agri-culture across north-central Europe. In this "complex adaptive system," households were its "grains of sand, its snowflakes, its flying geese." (Price 2000, 216–217)

COLONIZATION OR ACCULTURATION: EXPLAINING HOW AGRICULTURE SPREAD

The potential role of households brings us to the debate about how farming spread across Europe. Archaeologist Albert Ammerman and geneticist Luca Cavalli-Sforza reinvigorated this debate with a model they proposed in 1973, named the Wave of Advance. It describes the progression of agriculture across Europe in terms of colonization, from southeast to northwest, at an average rate of one kilometer per year. The authors named that part of the equation pertaining to colonization "demic diffusion"; very simply put, farming spread because farming peoples moved with their techniques, cultigens, and domesti-cated animals across the European landscape. Indigenous hunter-gatherers were likely either marginalized or rapidly amalgamated by farmers. At most, they played a limited role in the triumph of agriculture on the continent.

Demic diffusion has been questioned on both theoretical and archaeological grounds. Farmers may well have been the more vulnerable players, their settle-ments easy targets for mobile hunter-gatherers, and the ability of their small numbers to cut into the subsistence basis of foragers dubious at best. Strong evi-dence for continuity between Mesolithic and Neolithic communities has arisen from sites as far-flung as Ireland, western Scotland, Scandinavia, and eastern Europe, where skeletal and artifactual remains show that it was the foragers who, in what may have been a more gradual, selective, diverse fashion, *became* the farmers. Along the coasts, rivers, and lakeshores of Europe, Mesolithic peo-ples reminiscent of American Northwest Coast cultures had established semisedentary, complex societies—these were precisely the people most likely to adopt agriculture according to the social model discussed earlier. With north-central Europe and the southern Netherlands, Belgium, and the Paris basin ex-cepted, agriculture in our broad area of focus arose from a Mesolithic context. This record then calls into question notions of a coherent "Neolithic package," transmitted intact everywhere agriculture appeared. Marek Zvelebil proposes to "see the early Neolithic outside central and southeastern Europe as a period of gradual adoption of agro-pastoral farming with individual components of the Neolithic package (pottery, cultigens, domesticates) being adopted at different rates over a long period of time, and allowing for the adoption of these individual elements by the local hunter-gatherer communities." (Zvelebil 1989, 380)

Even where colonization most likely occurred, such as in north-central Europe, the more precise dates show that it did not spread at a steady pace. In fact, it took place about five times faster than predicted by the Wave of Advance model. A step model showing periods of rapid change followed by stasis seems to better fit the data.

This summary of current scholarship on the European Neolithic suggests key questions for environmental historians. First, if the European environment did not heavily determine the original adoption of agriculture, how then did the diverse ecosystems of northern Europe shape agropastoral systems? And second, if less colonization occurred than previously thought, then why did well-adapted Mesolithic peoples radically alter their subsistence systems? Though hardly uniform and arguably gradual, farming nonetheless began to change not only European social structures but also ecosystems. Thus, a third question arises: did agriculture begin to modify the nature of northern Europe? And finally, what cultural changes impinging on the relationship between people and nature accompanied agriculture? We will locate some responses to the first two questions in the context of four regional case studies—north-central Europe, the Baltic and North Sea coastal zones, Scandinavia, and the British Isles. A general discussion of the latter two questions will follow.

North-central Europe is defined by the sprawling midsection of the continent where Neolithic colonization most likely explains the transition to agriculture. Spreading from the middle Danube valley at around 5500 BCE, the LBK—the German acronym for Linear Pottery Culture—ultimately brought its technologies, plants, and animals north to the lower Oder and Vistula rivers, the Ukraine in the east, and the Paris basin in the west. The first farmers settled in the loess belt, whose easily drained, fine-grained soils were the product of glacial dust (see chapter 1). An upland area more homogeneous and more fertile than the North European Plain, the loess was irrigated by innumerable streams and small rivers. Dense deciduous forests grew throughout the region, though the driest loess areas may have had an open canopy or no forest at all. Hunter-gatherers had burned forests periodically for several millennia before the first farmers encountered them.

The hunter-gatherer populations were sparse on the ground, and agriculturally based communities proliferated throughout most of this region of 750,000 square kilometers in about one millennium. Interestingly, rapid diffusion has been used to argue for both native acculturation (migration could not have occurred as rapidly as agriculture spread) and colonization. Other evidence supporting the colonization hypothesis includes LBK culture itself; its pottery, house forms, stone tools, and domesticates show no local development within north-central Europe. Farmers may still have interacted with hunter-gatherers; archaeological evidence in the Paris basin suggests exchanges between the two groups of carbohydrates for animal protein, for example. The ultimate absorption of Mesolithic populations may have occurred through the siphoning off of adolescents as mates and laborers in farming communities.

A picture of early Neolithic farming in north-central Europe leaves much to conjecture, given the difficult recovery of plant residues and the poor survival of bones in loess soil. Certain elements do form a pattern. For example, the first farmers of the loess tended to settle along smaller streams, back from the banks where alluvial plain and watershed come together. Soils remained well nourished by both flooding and groundwater movement. The loess farmers selected from the repertoire of grains domesticated in various regions of the Middle East and transmitted via southeast Europe: they heavily favored emmer and einkorn wheat, cultivating barley, millet, and rye to lesser and varying degrees within the LBK region. Swidden agriculture, a general term indicating the shifting of fields through the landscape by burning new areas and allowing abandoned plots to return to scrub and forest, was assumed from the mid-1920s to be the earliest central European pattern, yet research since the 1970s has shown settlements of long duration and demonstrated the fertility of the loess. Now a model of horticulture, implying intensive cultivation on fixed plots, is most often applied to the LBK area. Horticulturalists were also herders; they hunted little if at all, raising mostly cattle but also some goats and sheep, and possibly producing cheese, as indicated by ceramic sieves.

By about 4400 BCE, central European farmers began to move beyond the loess in larger numbers, settling on the North European Plain or moving south to the Alpine Foreland. Though he came from the Italian side of the Alps, the "Iceman"—the world's best-preserved ice mummy, discovered in 1991—lived around 3300 BC and died at a high elevation on a glacier near the Austro-Italian border. Microscopic analysis showed that his last meal included unleavened bread made from einkorn, which had to have been cultivated. Only generations' worth of accumulated knowledge of the local climate, soils, and wild flora and fauna—a "cognitive mapping" that took place microregion by microregion—allowed expansion beyond the most fertile lowlands. Population growth and the development of more autonomous households may also have played significant roles. Once established on the North European Plain, farmers of the later Neolithic congregated in smaller settlements and relied more on wild animals and pigs, which were at best semidomesticated. These characteristics and the ap-

Mummy of Otzi the Iceman found in the Italian Otztal Alps in 1991. Various items were found with the mummified man, including a wooden backpack, a copper ax, a dagger, and a bow with arrows. (Corbis/Corbis Sygma)

pearance of small, round huts in Denmark, contrasting with the distinctive central-European longhouses, may indicate a melding of Mesolithic and Neolithic traditions.

In the Baltic and North Sea coastal zones, hunter-gatherers continued to prosper for over a millennium after the first appearance of farmers. The cultural groups of these regions—the Ellerbek of the North European Plain and the Ertbølle groups of Jutland and Zealand—exchanged materials with farming communities and thus knew of agriculture and its products: the two communities coexisted for fully 2,500 years on the Polish Plain. Thus the process of "becoming Neolithic" was slow and consciously rejected for a long period. These hunter-gatherers had grown more sedentary and populous than their counterparts in the loess belt, thanks in part to the munificent coastal environment. Communities fished more intensively, however, only toward the end of the Mesolithic as wild herbivores declined in number. They fished with impressive sets of gear, catching fish with hooks or spearing them from boats, and setting stream fences for mass capture as fish migrated in the spring and fall. Off the coasts of Norway, hunters harpooned seals and whales. The self-sufficient, complex culture of the late Mesolithic Baltic basin culminated during the latter part of the Atlantic climate phase, a high point in terms of temperature, precipitation, and sea level occurring around 3900 BCE. The agricultural frontier shifted north only when traditional resources became less reliable, as may have happened to the Ertbølle communities, who faced a shortage of oysters, a key element in the winter diet, due to falling sea levels. Foraging groups adopted agriculture, remaining in the majority even as they fused with Neolithic newcomers.

Likewise, the Rhine-Maas delta region was home to a similar story of cultural fusion. In this zone of numerous tidal flats and marshes, cultivation of wheat and barley likely began on floating peat islands, yet these quasi-farmers continued to exploit many species of fish and birds in addition to wild apples and berries, rosehips, and hazelnuts. Ultimately adopting pottery and animal husbandry as well, they kept their original settlement systems and social structures intact. Fully Neolithic groups encountering the delta may have scaled back their agricultural systems, "becoming Mesolithic" to a degree as they measured the lack of necessity to turn wholly to a farming economy. The northern Atlantic fringe represents one of the clearest cases of gradual and piecemeal adoption of parts of the Neolithic package in the context of stable populations and environments yielding a wide array of resources.

The geographic context of the Scandinavian Neolithic is of central importance: the northern two-thirds of Scandinavia are mostly uncultivable, even though the North Atlantic current has a tempering effect on the region. Coming into boreal Europe from the east, the earliest Neolithic is known as the Funnel Beaker Culture, a name denoting funnel-neck beakers, the most common pottery form, usually accompanied by evidence of cereal cultivation, livestock, and earthen long barrows. The transition to agriculture was both slow and rapid, depending on one's perspective. The agricultural frontier had stopped just 200 kilometers south of Scandinavia for over a thousand years, though foragers were obtaining pottery and other artifacts from the nearest farmers. Coastal and inland people also interacted, and both hunted and trapped a wide range of animals, including deer, wild pig, and fur-bearers. Then, at 4000 BCE, the Funnel Beaker Culture exploded across southern Scandinavia in an archaeological instant—less than 200 years. After that, however, no substantial forest clearance (key evidence of agriculture) took place for another half-millennium. Subsistence still depended on fishing and hunting terrestrial and marine mammals; cereal cultivation and herding only gradually assumed importance, earlier in Denmark and Scania, and later in adjacent areas. Even in the south, marked for-

The small winter hunting-fowling-fishing camp of ca. 4300 BCE at Bergschenhoek,
illustrating the continuity of the Mesolithic type of settlement systems, subsistence,
and perception of nature into the Early Neolithic in regions north of the loess.
(Republished with permission, Case Studies in European Prehistory, *Peter Bogucki, ed.,*
1993; permission conveyed through Copyright Clearance Center, Inc.)

est clearance, sturdy houses, megalithic tombs, and intensified trade in flint, copper, and amber did not appear before 3300 BCE.

The maintenance of a traditional subsistence base is one piece of evidence for the acculturation of Scandinavian hunter-gatherer-fishers. The animals and plants associated with agriculture appeared long after the pottery, polished flint axes, adzes, and bone combs; for some centuries the indigenous northern people remained indifferent to cereals, cattle, and so on. Moreover, aspects of material culture, namely the flint axes, show continuous development from the Mesolithic to the Neolithic. Funnel Beaker pottery is novel, yet the late Mesolithic Ertbølle produced ceramics, and some specialists locate developmental stages between the two pottery traditions. Finally, earliest Funnel Beaker pottery occurs at the latest Mesolithic sites, and skeletons differ only slightly between Mesolithic people and their immediate Neolithic successors. As T. Douglas Price summarizes, "Today, it is generally agreed that the last hunters of southern Scandinavia became the first farmers." (Price 2000, 293)

Some scholars point to environmental stress in the Scandinavian transition to agriculture. The onset of the Sub-Boreal climatic episode may have caused an increase in forest fires and erosion; dropping sea levels (more water became trapped in ice, as had happened during glaciations) could have reduced numbers of fish and marine mammals in the fjords and straits. This evidence remains contested, and there is little indication of declining population or health due to food shortages in the late Mesolithic. The argument based on prior social change, explained earlier in this chapter, may prove most relevant: as differences in status emerged among foragers, possibly due to contact with farmers, surplus production was needed to support accumulations of wealth, elite burials, and other ceremonial activities. As productive as it was, subsistence based on hunting, gathering, and fishing could not produce an adequate surplus, whereas agriculture could. Exchange networks widened—the demand for flint grew so keen that people in northwest Zealand began to recycle it—and came to incorporate exotic products such as copper ornaments issuing from central Germany or western Austria. The importance of the Neolithic might lie as much in new ideas about society as in the products and trading systems that embodied them.

In parallel with the Scandinavian case, most scholars of the British and Irish Neolithic argue for indigenous adoption of agriculture, although many remains disappeared after the submergence of Britain's Mesolithic coastlines. The paucity of evidence raises doubts about the habits of late hunter-gatherers (whether they were seasonally mobile or sedentary, for example) and thus about the reception of Neolithic novelties. Generalizing across Britain, Alisdair Whittle draws from isotope analysis of human bones to argue for a swift adoption of new, agriculturally derived foods at around 4000 BCE. He proposes a model of "filtered colonization" in which indigenous Britons acquired the domesticated animals and cultigens from a small group of newcomers. (Whittle 2002a, 91) Ideas and gifts may have circulated widely within a large zone of interaction in Britain and Ireland. Seafaring Britons may also have encountered Neolithic groups and their sheep, cereals, and pottery on the European mainland, bringing back samples and gradually integrating them.

Rather than replacing their old staples, Neolithic Britons cycled new ones into their repertoire without revolutionary change. Cereal growing was of minor importance for the first two millennia after its introduction, and early cultivation need not have tied down whole communities. Rather, crops may have been sown and left to fend for themselves, or watched over by only a portion of the community as the rest moved on. Similarly, sheep need not have occasioned immediate deforestation, for the many open landscapes, such as the southern chalk downlands, lent themselves well to grazing. In the uplands subsistence clearly revolved around pastoralism, yet cereal cultivation was to expand in Scotland in the early Bronze Age.

THE NEOLITHIC "MIND" AND
THE EFFECTS OF EARLY FARMING

If many inhabitants of northern Europe were selecting from and adapting to the Neolithic, rather than being displaced by it, then presumably they were also selecting from the diverse worldviews that had developed with farming. The adoption of agriculture did not revolutionize attitudes toward nature, though more casual accounts insist on the advent of linear thinking and a more alienated, domineering stance toward the natural world. Here as well, evidence is hard to come by. The best window onto the early Neolithic "mind" may lie in the hard stone of the megalithic monuments of Atlantic Europe. Whatever meanings they inscribe, however, these should not necessarily be associated with farming peoples in other regions of Europe. In areas bordering the Atlantic, from parts of Scandinavia to the Straits of Gibraltar, early Neolithic Europeans built earthen or timber mounds, passage graves, or stone rings during the fifth and fourth millennia BCE. Recent scholarship has rejected the notion that agriculture made such monuments possible, yet it has also turned to precisely the sort of question that interests environmental historians: What do the materials and settings of these monuments say about the visual encounter with nature?

Recent studies of Neolithic monuments in southern Britain, the Gulf of Morbihan in Brittany, and Galicia in Spain try to uncover references to landscape in the relationship between these monuments and their settings. These could include using natural features to echo human activities, such as burials; mimicking natural features, as one of the mounds at Petit Mont in Brittany mimics the shape of the headland; or using unfinished, local stone to deepen the relationship between monument and setting. Monument builders may have wanted to invest their structures with essences of the natural, "unlocking the sacred potential that was recognized to be present in natural landforms," or perhaps to highlight some feature "considered already to be present, immanent, in the landscape." (Scarre 2002, 10, 12) If the monuments seem to assert loudly their embeddedness in nature, that assertion may relate to an overall reckoning with the increased mastery over nature that came with agriculture. Through domesticating certain plants and animals, the Neolithic boldly called into question those species' relationship to human communities.

Monuments were related to sacred sites perhaps, but the megaliths almost certainly also had ritual significance within nature-based religions. One of the best-known megaliths is Stonehenge, built in phases on open chalk downlands in southern England. Completed with the addition of large sandstones around 2000 BCE, Stonehenge has elicited interpretations that have variously emphasized its evident monumental quality, its astronomical alignments, and its possible function as a temple for rituals of renewal or even for the re-creation of the recently dead as ancestors. Current thinking tends away from older views of Stonehenge as simply an astronomical observatory and toward an explanation of the alignments as a means of storing the sacred energy of the sun and moon. The alignments may also have been used to predict those days (such as solstices) when such energies would be most concentrated. Preceding Stonehenge by about two millennia, the recently discovered Goseck Circle in Germany shows a sophisticated ability to measure the sun's movements, but there too, astronomy appears to have been embedded in ritual, for the Goseck Circle was clearly also used for funerary rites or perhaps human sacrifice.

Whatever their precise relation to ecosystemic change, the monuments are synchronous with early Neolithic alterations to northern European ecosystems. The formation of agroecosystems was the greatest single ecological change in Europe's preindustrial human history. Whether cultivating plants, harvesting edible parts of plants, slaughtering animals, storing food, or controlling propagation, agricultural activities modify the ecosystems upon which they rely. As in other ecosystems, green plants capture solar energy, which then flows through food webs to animal organisms, gradually dissipating into uselessness (entropy). Plant growth requires water and nutrients, whether provided by natural processes or human agency. Agroecosystems, however, are highly simplified food chains and are inherently unstable. Humans typically manage agroecosystems to favor a few annual plant species (cereal grasses in the case of Europe), subsidizing their growth with nutrients and water to gain high productivity (though less than the total productivity of the diverse species in a natural ecosystem). These plants concentrate their entire growth and reproductive cycle into high-speed circulation of nutrients during a single growing season. Because in nature such pioneer organisms soon give way to other, longer-lived ones, agricultural work is required to keep the system at this simple stage. These systems are unstable, both for their requiring subsidies and for their exporting of the favored consumable biomass to humans elsewhere.

More broadly, farmers alter vegetation by clearing forest and controlling weeds; they alter the composition of topsoil by hoeing or plowing; they drain

View of Stonehenge, Neolithic monument in southern England. Stonehenge likely had ritual meaning within a nature-based religion. (Corel Corporation)

and divert water; they retain seeds for further planting; and by selecting for desirable characteristics, they often render certain plants unable to reproduce on their own. Such changes may be significant at the microlevel, but for the early and middle Neolithic of northern Europe we must apply a wider lens.

Neolithic farmers did not fully re-create the mosaic landscape bequeathed to them by the Mesolithic, though they did clear more and larger openings in the forest with polished stone axes and by ring-barking and burning, leaving larger stumps to rot. The composition of the ancient wildwood remained intact. Far from being an unmitigated enemy, the forest provided pasturage for pigs and cattle, the latter grazing on edge grasses and munching on twigs, leaves, and young saplings. Ancient Britons practiced coppicing, allowing the cut stump of a deciduous tree to send up shoots used for cattle food, poles, and other purposes. The forest became more open, with increasing edge, clearings, and flourishing understory. One can imagine the small fields ensconced in woods that little by little developed into an agrarian landscape. Deer populations may have taken off as forest edge increased, hunting declined, and farmers began to kill wolves and other predators of both deer and livestock. Neolithic populations remained too small to take much of a toll on either forest or soil. No evidence of decreasing soil fertility comes from central Europe, though the earliest farmers in the region relied too heavily on cattle, an imbalance corrected in later prehistory by a partial shift from cattle to domesticated pigs. In this light it is best to see the Neolithic as the first stab at a long work-in-progress.

Environmental change accelerated in succeeding centuries. From the creation of landforms, such as some of the lowland heaths of Britain, to the podzolization of soil in Denmark, human communities began to rework ecosystems as agriculture expanded. Yet these centuries are named after the metallurgical industries that drew only small amounts of labor away from agriculture and had only very localized impacts on ecosystems. Early metallurgy deserves attention, however, in that it represents the incorporation of an additional subset of nature, metallic ores, into human orbits of production and consumption; it also helped link far-flung regions of Europe together through trade.

BRONZE, IRON, AND AGRARIAN SOCIETY

Artisans in eastern Europe may have been working copper from as early as 4500 BCE, but in central and western Europe they first cast bronze around 1850 BCE; the Bronze Age ends at about 750 BCE. Smiths tinkered toward the classic blend of 90 percent/10 percent alloy of copper and tin by first adding small amounts of antimony, arsenic, silver, or nickel to copper. As they perfected the alloy, they also developed methods of heating in kilns and casting in two-part molds. Bronze artifacts, ranging from neck rings and bracelets to swords, daggers, axes, and ultimately sickles, were prized over stone items for the greater flexibility in size and shape allowed by molten metal. Although metal is softer than stone, the added tin increased copper's hardness. However, few people possessed metallurgical skills, and the raw materials were less ubiquitous than stone. Common in Spain, England, Ireland, mountainous central Europe, and the Urals, copper is nonexistent along the North European Plain and Scandinavia, and rare in France. Tin ores lay concentrated in Brittany, Galicia, southwest Britain, northwest Italy, and Bohemia. Trade in ores grew brisk as demand spread across all of Europe, yet the relative rarity of manufactured bronze guaranteed that the items mentioned above remained in the hands of elites. Hoards of weapons and tools, in a few cases containing thousands of pieces, testify to concentrations of wealth not seen in the Neolithic.

Due to the rarity of copper ores, mining left gashes and large galleries in only a few places. By 1700–1500 BCE, surface ores in the major mining region around Salzburg, Austria, had been exhausted, and miners worked deeper into hillsides by means of shafts driven in up to 100 meters. According to one esti-

Illustration of men working at a Bronze Age foundry and workshop, ca.1200 BCE. (Getty Images)

mate, the thirty-two mines surrounding the Mitterberg lode became exhausted in only seven years. Perhaps 13,000 metric tons of copper ore were extracted, the operation consuming 20 cubic meters of timber per day for props. Smelting also required trees in the form of wood and charcoal to heat the hearths and kilns; though far from the greatest cause of deforestation in ancient northern Europe, cutting for the metallurgical industries could take its toll on forests, a toll that would increase in the Roman Era.

The next metallurgical breakthrough—iron smelting—promised greater fuel economy than bronze allowed, but iron implements drew their popularity from the metal's superior strength, durability, and capacity to keep a good cutting edge. Iron ore could be found throughout Europe close to the earth's surface. It did not require alloying, although smiths mastered only gradually the techniques of forging at high temperatures. Between 1000 BCE and 750 BCE iron metallurgy became known to most of Europe south of Scandinavia, more quickly than either agriculture or bronze technology had spread. Europeans retained bronze for vessels and ornaments, but now used iron for tools and particularly weapons. Modern weapons were iron weapons, and with them a culture

Illustration of an ancient Celtic encampment. (Stapleton Collection/Corbis)

of mounted warriors came to characterize the elites of northern Europe. More available than bronze, iron diversified and democratized to a degree, becoming the material of choice for axes, augurs, and even nails. The Celts, a people of west-central Europe who emerged around 500 BCE, used ploughs with iron shares by the first century BCE. The pre-Roman Iron Age north of the Alps dates from about 750 BCE to 1 CE, the latter year a convenient date acknowledging the beginning of the Roman Era. It, too, was surely an age of iron, hence the expression "Roman Iron Age" (which applies beyond the borders of the Empire). In terms of both agriculture and metallurgy, "barbarian" Europe was as sophisticated as Rome into the first century BCE. Caesar himself noted the iron chains used by the Veneti, a Gallic people based in southern Brittany, to attach anchors to ships, at a time when the Romans still used rope.

If metallurgy and its products were the novelties of the second millennium BCE and beyond, the subsistence base did not remain static. Northern European society turned fully and irrevocably to agriculture at this time, following the pioneers of the Neolithic. More settlements, signs of greater territoriality, newly domesticated species, regional specialization—these and other changes betoken the foundation of an agrarian society that would endure for over two thousand years, until the introduction of potatoes, maize, and tomatoes from the Americas in the sixteenth and seventeenth centuries CE. In some respects, it lasted up to the Industrial Revolution.

A glance at Britain alone shows a startling evolution from the early Bronze Age: the "monumentalization" of landscape by means of the megalithic stone circles and barrows, discussed earlier, had given way to boundaries by 1500 BCE. Britons laid banks and ditches across extensive tracts of land, delimiting rectangular fields, traces of which persist. Motivations behind this trend toward circumscribing the land remain opaque and cannot have derived only from population growth. These so-called Celtic fields were extensive in northwestern Europe, and they may reflect intentions to improve the soil: banks helped prevent erosion, and enclosures facilitated manuring. Many of the fields may have been reserved for grazing, although their typical size—the largest were no more than a half-hectare—suggests fields that could be ploughed in a day. In Denmark, too, farmers laid out very regular fields and most likely enclosed them at the beginning of the Scandinavian Iron Age, 500 BCE. Though the existence of such enclosures elsewhere on the European continent remains an open question, methods of land improvement may well have been practiced generally.

The Bronze and Iron ages were occasionally times of agricultural surplus, as northern Europe exported grain and salted meat to the Mediterranean in exchange for luxury goods known since the heyday of Greek trade in southern Europe. Beyond soil improvement, advances in harvesting technology and animal husbandry made surpluses possible. Bronze tools followed by iron sickles and later iron scythes found their way into the fields and meadows. Hunting declined steadily in Europe, while reliance on livestock intensified. Cows were used increasingly for dairying and draught purposes; herds were more effectively controlled through castration, their longevity improved by winter housing. Pigs gained importance in Poland and the north of France, and migrants brought wool-bearing sheep from the Middle East. The horse, domesticated in the southern Ukraine in the centuries preceding 3000 BCE, had come to central and western Europe with pastoral peoples migrating west in the Late Neolithic. The work of horses heightened the efficiency of traction and transportation, and horse meat and mare's milk contributed to human diets. Celtic and Germanic peoples revered horses, and Epona, the Celtic horse goddess, protected both horses and the people who worked with them. Horses played an increasing role in warfare, and the Romans would later admire the horsemanship of Rhineland Germans.

A surviving paved Roman road in England. (Robert Estall/Corbis)

Subsistence and surplus issued from agroecosystems that were organized as follows. The *ager*, or cultivated fields, could be permanent or temporary, and larger livestock would have cyclical access to them. The *saltus* of "waste" or scrub, lying beyond the fields, provided the key grazing resources, and the out-lying *silva*, or woodland, provided raw materials, wild plants, grazing for larger animals in the more open areas, and habitat for pigs. Within this general scheme, diversity reigned across Europe, not the least element of which was the increasing array of cultivated plants. Moving away from the nearly exclusive reliance on Anatolian wheat varieties during the Neolithic, Bronze Age farmers blended barley into their repertoire in Britain, the Netherlands, and much of Germany; spelt later made its appearance in Switzerland, southern Germany, and Britain (see below), and rye moved west out of eastern Europe. Since not all of these cereals need be planted in the same season, Bronze Age farmers probably introduced crop rotation. Rotation may have been responsible, along with shorter fallow periods, for the long-lived villages along the shores of Lake Clairvaux in Switzerland. Exploitation of marine resources in Scandinavia and along the Atlantic littoral continued much as during the Mesolithic and Neolithic, and Scandinavians continued to make full use of wild plants. Toward the end of the period covered by this chapter, a man in Denmark fell to a ritual killing; discovered in a peat bog and examined in the twentieth century, his well-preserved stomach contents showed a last meal of gruel composed of the seeds of sixty-six plants, all but five of them wild.

Not all agricultural villages were as successful as those along Lake Clairvaux. Periodic subsistence crises mark European prehistory; they resulted in dramatic alterations to the landscape as entire villages relocated, and forest or scrub reclaimed abandoned fields. The light soils of the North European Plain could not, for example, stand up to either more intensive cultivation or greater concentrations of livestock. Communities in early and mid–Iron Age Denmark cleared wood pasture in favor of permanent meadows, created in tandem with continuously cultivated fields. Thus a system of intensive agriculture replaced the old extensive practices. One key problem lay in providing enough winter fodder for animals, now housed in sheds throughout the winter. Animals normally ate their way through harvested hay in the course of winter, yet the drier centuries of the late Iron Age brought less grass. Deforestation (thus the destruction of permanent fodder) called for larger pastures, requiring in turn more deforestation in a dangerous spiral. A highpoint of forest clearance was reached in Denmark by the middle of the Iron Age; the open landscape then receded, not to be attained to the same degree until 1300 CE.

In Britain, too, the bounded land contained far fewer trees; many forests either became sparser wood pasture or were transformed into cultivated fields, some of which later evolved into moorland and blanket peat. As much as half of the forest of Britain present in the late Mesolithic had disappeared by the Iron Age, with the deforested highlands contrasting with more wooded lowlands. Animals dependent on forest declined as well, with beaver, bear, and wild boar populations dropping and wild cow going extinct in Britain by about 1350 BCE. Forests succumbed noticeably to fire and sharp metal axes even in early historic France, where dense forests may have been more legendary than real; some authors claim that France, particularly the Paris basin, center, and west, was dotted by small woods and copses by Caesar's time. The general's troops faced a thickly wooded barrier only in the Ardennes. Sheep, so often agents of deforestation, began to dominate livestock herds in the center-west and south of France. Caesar reported the existence of aurochs in the Hercynian Forest, which stretched across southern Germany into Romania; if true, then the presence of these grazers signals much open grassland, not a closed forest.

The major historical event of pre-Roman northern Europe might reflect a degree of ecological instability. From the middle of the fifth century BCE, migrants in the tens of thousands streamed to the south and east from the Marne-

Sculptures of mother goddesses from the Rhineland during the Early Roman Period, 164 CE.
As shown here, mother goddesses were frequently portrayed with fruits, evoking fertility
and abundance. Altar of the Matron, Rheinische Landesmuseum, Bonn, Germany.
(Art Archive/Rheinische Landesmuseum Bonn/Dagli Orti)

Moselle region—the heart of the La Tène culture of the Celts. New warrior
elites were on the rise, and the migrations they led involved conquest, the
spread of Celtic culture, and the resettlement of Celts themselves. One stream
headed south, penetrating the fertile valleys of Italy by the end of the fourth
century at the latest, and another stream moved east along the Danube, eventually settling largely in the Carpathians and the Transylvanian Alps. The image
of restless Celtic warriors seeking plunder belies the full story: some evidence
suggests a "major systems collapse" brought about in part by population growth
that overshot the capacity of west-central European agroecosystems. Ancient
writers such as Pompeius Trogus and Livy believed that the Gauls had "out-

grown their land"; the former estimated that 300,000 people had packed up in search of new territory to farm. (Cunliffe 1994, 358–361) Peter Wells cautions, however, that archaeological evidence does not support these historical sources; Celtic settlements and cemeteries show no signs of large-scale, permanent out-migration. In other words, numerous migrants clearly did leave, but many may have returned later, as various historical migrant communities have been known to do. (Wells 1999, 46)

For Bronze Age and Iron Age Europeans, the natural world furnished not only the means of subsistence but also the elements of religion and ritual. Numerous deities inhabited their world and lent a sacred aura to specific aspects of nature. The Celtic goddess Sirona, for example, was linked to the healing properties of warm springs, as was Apollo-Grannus (a deity of mixed indigenous and Roman origins), to whom a temple was dedicated in western Bavaria in about 160 CE. Trees, especially the sacred oak, held important symbolic value; worship of deities took place in sacred groves, where hunting and gathering firewood were proscribed. Waterways and pits received offerings to deities, sometimes on a massive scale. A major site of sacrificial offerings is Gournay-sur-Aronde in northern France, where for three centuries prior to the Roman conquest hundreds of thousands of animals met sacrificial deaths. Weapons, too, were destroyed and deposited there, suggesting perceived links between natural and human powers. As much of our knowledge of northern European religions comes from Roman sources, it is to the Roman Era that we now turn.

NORTHERN EUROPE ON THE PERIPHERY: THE ENVIRONMENTAL LOGIC OF THE ROMAN EMPIRE

The Roman conquest of Italy proceeded gradually in the course of the third century BCE; Romans ultimately expelled the Boii, a Celtic tribe, from the Po Valley in the campaigns of 196 and 197 BCE. Spain came under Roman rule over the first three-quarters of the second century. Rome extended political domination to northern Europe in the two centuries flanking the beginning of the common era, with Caesar's conquest of Gaul in 58–51 BCE and the Claudian invasion of England in 43 CE. Military campaigns east of the Rhine, undertaken in hopes of launching an invasion of Bohemia, were abandoned after crushing defeats at the hands of Germanic tribes under the warrior Arminius, who used the forest to effect in annihilating three Roman legions. After 16 CE the Emperor Tiberius withdrew troops to the Rhine, which remained a heavily garrisoned frontier until the late third century CE. Thus, the North European Plain, Scandinavia, and most of central Europe lay beyond the Upper Germanic Limes (the Rhine boundary) and the Rhaetian Limes (the Danube boundary) throughout the history of the empire.

Roman influence hardly stopped at the Limes, however: the economic reach of the empire extended far beyond military frontiers. The Roman trading system would transform even the economies of Scandinavia and the North Sea region, and market economies grew up in the frontier zones themselves. To take the more familiar example of Gaul, Roman merchants were anchoring their ships at its southern ports and transshipping cargoes, largely of wine in bulky amphorae, northwest along the Aude-Garonne river route as well as north along the Rhône River long before Caesar's conquest. They were in fact tapping into, and deepening, the well-established trade routes of Iron Age France. Renewed Germanic invasions over several decades of the first century BCE threatened Gaul more than Italy and provided Caesar with his rationale for conquest. By then southern Gaul was especially Romanized, but trade in wine and consumer durables such as tableware was also brisk in Brittany and across the Channel. On the other side of the Rhine frontier, a trading zone extended for some 200 kilometers in which Germans avidly sought Roman coins to purchase commodities, and where both Romans and Germans acted as commercial middlemen.

Northern Europe formed, to varying degrees, much of the "barbarian periphery" upon which the "Roman Mediterranean core" relied for raw materials, slaves, markets for excess wine and other goods, and settlement grounds for retired soldiers. (Cunliffe 1994, 414) The terms "core" and "periphery," at the heart of world systems theory, oversimplify imperial relationships when analyzed on a regional basis, for provinces carried on lively trade with one another that was taxed at only 2 percent. Nonetheless, these concepts provide a general framework for understanding the economic and environmental changes that accelerated under the Romans' political and commercial empire.

Extending across much of the northern periphery, for example, were the *oppida* constructed from the second century BCE. Large enclosed settlements varying greatly in size (they could range from 10 to 500 hectares), they emerged across temperate Europe, from western France to the Sudeten Mountains. The oppida marked a significant change in the settlement system of the continent, for even though most people continued to inhabit unenclosed villages, oppida introduced a distinction between those who lived within, and those who lived outside, the walls. Those walls became, as Peter Wells notes, "the largest human-built structures in the landscape of pre-Roman Europe." (Wells 2001, 85) Given their placement in areas where topography aided in defense, protection against invaders seems to have been their primary purpose, though ritual purposes may well have been at play, too. Paradoxically, in an age of quickening trade and over-all mobility, the oppida may have fostered, for their residents, a deeper identification with a specific territory as these centers acquired reputations.

The oppida were indigenous centers of artisanal production, housing urban economies that were flourishing prior to Roman commercial dominance. Impressive in both scale and diversity, handicraft production ranged from pottery and glass beads to ironwork and coins. These artifacts, which were destined to be traded, have turned up in large quantities at the well-excavated sites, such as Manching in southern Germany. Other centers such as forts and unenclosed towns developed during this period; they all bear the impact of money-based economies and an emerging nonagricultural sector. Though Roman merchants were heavily involved in the export of goods manufactured in the Roman core—the pottery, glass, gems, and textiles increasingly in demand in the north—trade in goods from the periphery also took place, in addition to the agricultural and forest products that changed hands in the oppida. Above all, the advent of formal empire meant significant levies and taxes that necessitated more commerce so that payment in silver coin could be made. The three provinces of Gaul figured among an inner ring of provinces that paid taxes in cash and were net exporters to Rome.

Rome's fiscal demands, in addition to early urbanization from the Iron Age onward, inevitably affected northern European agriculture. No longer just a hedge against bad harvests, agricultural surplus was now necessary to feed the larger numbers of nonfarmers and to pay the taxes exacted by Rome. In frontier zones where Roman legions were stationed, officials usually collected taxes in the form of agricultural produce. Careful studies of Britain show how agriculture expanded in the northernmost province of the Roman Empire. New crops now complemented the land clearances, improvements, and field boundaries of the Bronze and Iron ages. First spelt gradually supplanted emmer wheat; more hardy and able to grow in marginal soils, spelt allowed farmers to bring less desirable land, such as the clay areas of Suffolk, under cultivation and to extend the growing season. A second switch, this time to modern bread-wheat with its easily removable chaff, may have occurred in the early post-Roman period. Granaries and large mills remaining on the landscape of Roman Britain also provide evidence for larger surpluses, though Roman garrisons were also importing some of their grain. Archaeologists have found the seeds of exotic food items such as grapes, figs, olives, dates, pine nuts, cucumber, and coriander at Romano-British sites. Their presence largely signifies long-distance trade (though vine cultivation did spread to Britain) and a growing taste for "Roman" foods.

In contrast to southern Gaul, given over to slave-based cultivation of vines and olives, Britain saw a less radical transformation of agriculture. There, communities and individuals probably made the decisions to respond to higher demand. British farmers engaged in other new practices as well, producing malt and withholding some of the cereal crop for beer production, and intensifying cultivation in the form of market gardens that yielded vegetables and fruits. Farming was, on the whole, efficient enough to meet demand with relatively little reliance on either wild animals or the rich coastal resources of Britain. Although farmers in Britain as in the Netherlands did cultivate some coastal marshlands, the former and future wilderness of northwest Europe, the reclamation of coastal marshes occurred rarely. Pressure to produce a surplus justified the investment in flood defenses only in parts of the Severn Estuary of southwest Britain.

The Roman Empire produced no single economic or ecological effect on locales, though many underwent transformation in the ways suggested above. At the most general level, regionalization carved out distinctions within larger units of territory; for example, southeastern Britain, a region of villas, oppida, and rapidly expanding agriculture, emerged as the most advantaged part of the island under Roman tutelage. From the Rhine frontier, separate zones rippled toward the northeast, each marked by varying degrees of commodity production and exchange as an inverse function of distance from the frontier. A regionalized economy forged by empire thus led to greater regionalization of land use.

The direct ecological effects of the Roman imperial economy can be best suggested by examining traces of the following activities: road building, mining, and logging. Though ecosystems were being marred and natural capital was heading toward exhaustion in the Roman Era, northern Europe, though not immune from these effects, remained on the margins of ecological decline.

The roads that tied the empire together, including 90,000 kilometers of highway and over 300,000 kilometers of secondary roads, left gashes in the landscape that remained for centuries or longer. Roman engineers usually laid out roads as straight as possible, slicing through fields, boundaries, and villages. Deep as well as straight, the Roman roads required tunnels and embankments that disfigured landscape, much like the railroads of the nineteenth and twentieth centuries. Leading not only to Rome, roads crisscrossed the empire, facilitating the movement of travelers, officials, and postal servants, whereas freight traveled more often by river or sea.

Mining depleted resources and caused severe pollution. We can measure these assertions by the thousands of tons of ore extracted to produce pure metal, the trees required to prop up underground shafts and galleries and to smelt the ores, and the slag heaps measuring in the millions of tons, as in central Europe and Spain. The scale of preindustrial metal extraction could be immense: Roman levels of both copper and lead production, for example, were not matched in Europe until the Industrial Revolution. Particulates from both metals reached the troposphere over the Arctic. Beyond air pollution caused by smelting, notoriously high levels of lead poisoned bodies throughout the Roman Empire, from Asia Minor to England. Ubiquitous lead pipes, but more importantly lead-glazed containers and vessels and larger, lead-lined vats and cisterns, delivered lead to human bodies via water, wine, and food. Lead even found its way into food preservatives, sauces, and medicines. Roman aristocrats, heavy consumers of wine, ingested an estimated mean of 250 micrograms of lead per day, a level far exceeding the recommended maximum of 43 micrograms set by the World Health Organization.

Northern Europe did not suffer the catastrophic deforestation of the Mediterranean basin. J. Donald Hughes suggests a figure of 50–70 million acres of trees in the latter region cut down—for purposes of mining alone! (Hughes 1994, 126) Wood had manifold uses as fuel for cooking, heating, smelting, and manufacturing, and as a principal building material for entire towns, forts, and navies. Roman innovations such as central heating and public baths required enormous quantities of wood. By the first century CE, some ceramics factories had been relocated from Italy to southern Gaul due to lack of fuel in the Roman heartland. These demands for fuel and material must be viewed alongside the greater motive, in the ancient world, to clear the land of trees in order to farm it. Such was the fate of north Africa, transformed into the deforested bread basket of the Roman Empire. Even so, the wholesale consumption of timber inevitably told upon the denser forests of the north. Romans turned, for example, to the forests of the Vosges Mountains to supply wood for metallurgy and glass manufacturing, and they established multiple ironworks in both Gaul and Britain. Estimates vary widely as to the area of forest needed to supply fuel for a given output of iron. A conservative estimate based on the ironworks of the Weald (a word from Old English meaning "forest," here designating an area in southeastern England) shows that Roman iron-smelting relied on regrowth from coppicing after an initial phase, the industry thus settling into a more or less sustainable exploitation of forest. (Rackham 1980, 107–108)

Wood shortages could occur locally or regionally; they were beginning to afflict Gaul, already highly cleared in parts by the time of conquest, as suggested earlier. In the 200 years between 50 BCE and 150 CE, Roman-occupied southern Germany experienced significant deforestation, though pollen analysis indicates a cyclical pattern dating from at least 3000 BCE. No northern cities approached the population levels of Rome, Antioch, Carthage, or Alexandria, yet the large towns of London and Paris and the Rhineland colonies of Cologne, Strasbourg, and Mainz were, to name but a few urban areas, largely constructed of wood. So, too, were the forts and watchtowers built in increasing numbers along the Rhine and Danube frontiers, until Visigoths crossed the Danube in 375 CE, and Vandals, Alans, and Suebi overran the Rhine frontier early in the fifth century. In all, however, the imprint of neither population nor technology was deep enough to cause irrevocable degradation of northern European forests in the Roman period, quite unlike the situation in north Africa. The Romans did enact laws regulating cutting and protecting tree plantations in areas where shortages had become acute. Other regulations pertained to building practices for purposes of economizing wood in construction and capturing solar energy.

Finally, the twin perils of Mediterranean agriculture—soil erosion and salinization due to irrigation—were nearly nonissues in northern Europe. Here, denser soils, more forest cover, and greater precipitation played the conserving roles that they had before the Roman period and continued to do thereafter. Of larger legacy are the agrarian systems put into place imperceptibly since the Neolithic. Expanding from the Bronze Age and becoming more commercialized under the Romans, these systems survived the social chaos of imperial decline and successive invasions, contracting somewhat but remaining to provide the essential foundation of European society in the Middle Ages.

3

MEDIEVAL CHRISTENDOM IN GOD'S CREATION
Environmental Continuities, Coevolutions, and Changes

In general European history, the term "Middle Ages" commonly refers to the millennium between the fourth to sixth centuries CE, when there was a breakdown of a Mediterranean-based, Roman political and social order, and the late fourteenth through sixteenth centuries CE, which saw the emergence of a self-consciously distinctive "modernity." Despite such generously hazy boundaries, this retrospective label based on traditional political and intellectual history poorly fits the observed chronology of major economic and social developments that affected interactions between Europeans and their evolving natural environment. Even the great global environmental event that did coincide with the end of the Middle Ages, namely the new intercontinental contacts and exchanges initiated by the first voyage of Columbus in 1492–1493, only quite slowly and marginally exerted environmental influence in Europe itself.

Medieval European civilization is best defined by its eventually self-conscious adherence to forms of Christian beliefs and institutions derived from late Roman practice, conducted primarily through use of the Latin language (long the only important literate medium in the West), and acknowledging the bishop of Rome (the pope), successor to the Apostle Peter, as the leading religious authority (the meaning of that recognition varied greatly through the medieval centuries). Christianity began in the Middle Ages as an ideology largely confined to onetime subjects of the disintegrating Roman Empire, yet was still huge in its cultural magnetism. But by around 800 CE Christianity became the identifier of ruling elites all over the West, and by 1200 it was the touchstone of community membership for nearly all their subjects and northern and eastern neighbors from Scandinavia to the plains and steppes of eastern Europe. Western or Latin Christendom thus stood distinct from its sibling cultures, the fellow heirs to classical Mediterranean civilization in the Orthodox Christian East (Byzantium, the Balkans, Russia) and in the world of Islam (which at times during the Middle Ages included important parts of the Iberian and Balkan peninsulas and the insular Mediterranean region).

Throughout the Middle Ages, not only strong cultural allegiances but also important political and economic ties to dynamic leading centers in Italy, southern France, and Spain denied the separation of northern from Mediterranean Europe. Both fundamental assumptions about relations of humankind to nonhuman nature and such key material phenomena as demographic trends and crises, urbanization, and long-distance trade united both sides of the Alps into a single, medieval European civilization. Indeed, until well into early modern times, one continuing theme of European environmental history is the testing and adaptation to northern circumstances of cultural and technological innovations from the south.

CONTEXTS: POWER, WEALTH, AND MINDS

Medieval European society was fundamentally and unabashedly hierarchical, organized and conceived on the principle that some people were inherently superior to others. There were what we now can call elites, who were very few in number, and there were large numbers, up to and beyond 95 percent of the population, of what some medievals called "commoners." Elites possessed, in varying degrees, wealth, power, and prestige; ordinary people did not. Vertical gradations within elites recognized the distinct status (estate, rank) and rights of each. A monarchic idea, derived equally from Romano-Christian imperial principles and the warlords of Germanic peoples, set a king at the social apex, with authority both supported and limited by hereditary aristocrats (nobles) of land and family. Over the centuries various others (warrior knights, churchmen, top servants to rulers, lawyers, big merchants) eventually also gained relatively secure access to resources, influence, and rank. "Politics" and something like class consciousness were affairs of the elite. Solidarities both vertical (lordship, vassalage, clientship, political allegiance) and horizontal (kin, community, corporation) fueled their rivalries and helped resolve disputes. Medieval elites did not "work" for a living.

In material terms, medieval elites lived off the production of the commoner majority, though the forms of exploitation differed considerably. In fact, if not always in law, the lives and livelihoods of medieval commoners varied greatly in time, space, and stratification. Most generally it is true that the earlier Middle Ages, and for a longer period on the cultural and political margins, recognized a sharp polarity between free commoners, who were obliged only to public authority or a freely accepted lord, and definitively unfree slaves, who were the personal property of others. Trends for most of the medieval period then blurred this dichotomy into a gradient of liberties and private obligations held by nearly all. The generic modern image of a "half-free" medieval serf covers a great variety of conditions and limitations, legal, social, and economic, the latter including security of access to land, control of one's own labor, and the ability to make decisions affecting production and its environmental consequences.

By present-day standards medieval Europe was an impoverished society, controlling only small amounts of material goods. As today, people in such circumstances had to devote a very high majority of their economic activity to meeting their subsistence needs for food, clothing, and shelter with the natural organic products from advanced solar agriculture. In other words, most medieval Europeans were "peasants," that is, subordinated agricultural producers, working in household units of people and resources (family farms) to meet their own needs and to convey a surplus to support members of the elite and a few nonagricultural hangers-on. Gradual or abrupt shifts in lord-peasant relations are key phenomena in the medieval economic and social history of many regions, and in some instances they directly or indirectly affected the environmental impact of agricultural practice. While nonagricultural producers (metal, textile, or construction workers; traders; and others) were never absent, their economic role and social position gained importance only after the revival and spread in northern Europe of urban centers and of a market-based economic sector after 1000 CE.

The decline in human numbers characteristic of the late Roman Iron Age continued until around 600 or 700 CE, followed by two or three centuries of erratic or contradictory local trends. But by around 1000 CE evidence of denser settlement, expanding farmland, and the like indicates that growth had become general and was spreading from west to east. In most areas this trend accelerated into the thirteenth century. Western European populations likely peaked between 1300 and 1340 (regional data is disputed), while further east the upward trend may have lasted somewhat longer. Famine, epidemic disease, and warfare precipitated a sharp, mid-fourteenth-century decline that lasted well into the 1400s. Renewed growth thereafter probably regained early-fourteenth-century population levels by the mid-sixteenth century, though with distinctive regional differences and higher levels of urbanization. Europe's medieval demographic experience of trough, expansion, crest, and crash is central to its environmental history.

Thoughtful early-twenty-first-century observers of humans in relation to natural systems commonly think in terms of networks or webs of interconnections. Ecology is a postmedieval concept referring to the interdependence of living organisms and their environment, living and nonliving. To modern minds this necessarily includes humans in relation to nonhuman nature. Such mental connections differ from medieval thinking, which normally separated God from His Creation and man (*homo*) from the rest of physical Creation, while arranging all into a "great chain of being," a hierarchy from the highest and most spiritual (God) to the most base, material, and evil (a stone, Satan). What Christian Aristotelian philosopher and theologian Thomas Aquinas (1225–1274), a Neapolitan who spent most of his career in Paris, called the *connexio rerum* (gradation between related *kinds* of things) merely softened the division between the material and the spiritual that humankind straddled. More popular beliefs, even those that were officially proscribed, populated the natural world with a whole array of supernatural beings, saints, demons, angels, and occult powers, which the prudent, brave, or foolhardy might seek to manipulate through appropriate or prohibited rites. A magical view of the universe leavened the orthodoxy of official medieval Christianity.

The Christian theology generally accepted during the Middle Ages recognized the whole physical universe as created from nothing and subsequently sustained only by the divine will. God created humankind in His image and placed them on earth to govern all Creation. But the consequences for environmental relations of that doctrine (and mankind's subsequent fall into sin and separation from God) are by no means self-evident. Medieval thinkers did assume human use of animals, plants, and inanimate nature as enjoined by God but, insofar as they bothered to engage these issues at all, they dwelt more on the injunction to "increase and multiply" than on any implications of "dominion." Writers who favored sexual abstinence on spiritual grounds hedged the command to "multiply," as did a few late-thirteenth-century authors who seem dimly aware of population pressures in relation to landed resources of their own time.

Historian David Herlihy has labelled "apocalyptic" such negative medieval attitudes toward the environment and contrasted other "adversarial," "collaborative," and "recreational" positions among medieval thinkers. Fear of natural forces could be expressed in tales of monstrous beings, like the fearsome Grendel and fierce dragon in the Anglo-Saxon epic *Beowulf*, or in a dichotomy between civilized Christian society and an "other" variously expressed as wilderness (*deserta*, the "desert" of Christian scripture) or the forest (as in Arthurian romance literature). But just as later the courtly knight found in the forest the adventures that validated his civility, early medieval Christian writers knew a certain ambiguity toward the "desert," where the first Christian ascetics, Egyptian hermits, had fled sinful human society to test themselves against the wiles of the devil and their own flesh.

The Altar of St. Bernard painted ca. 1500 by Joerg Breu the Elder (1473–1537) at the Cistercian monastery of Zwettl, Austria, depicts the monastic brethren harvesting cereals with the traditional sickle. (Erich Lessing/Art Resource)

Since at least the seminal text of Western monasticism, the mid-fifth-century *Rule* by the Italian Benedict of Nursia, the much-respected monks placed positive value on labor (both manual and intellectual) and then recognized the human role in landscape change. They thus saw themselves as laboring to make of the desert a "paradise" (literally, an enclosed garden, so symbolically representing the monastic precinct itself). Only a different, nonmedieval sensibility sees the irony that monks, who sought "wilderness" to escape social entanglements, then built their "paradise" by humanizing the landscape. Monastic culture so valued the process of turning wilderness into paradise that later monastic foundation legends, too often taken literally by an earlier generation of modern historians, even constructed mythic wilderness in locations later researchers have shown were subject to prior human use. Some medieval contemporaries did recognize a further irony, when ideologically driven monastic founders in the central Middle Ages expelled whole peasant communities to create their own "desert." Still, the surviving written record does show monks leading agricultural colonization in certain woodland regions, such as the early medieval Vosges or the twelfth-century Polish borderlands. Over a very long period of time, the dedication to self-sufficient solitude for spiritual ends on the part of long-lasting Cistercian (later Trappist) and Carthusian houses helped create and preserve some of Europe's most sustainable, and now most idyllic, humanized landscapes.

By the twelfth century, human work in Nature, initially conceived in an adversarial sense, was more often also portrayed in collaborative terms. Clerical intellectuals like Bernard Sylvester (ca. 1140) and Gilbert of Poitiers (de la Porée, 1075–1154) saw humans working with God's Creation to complete the divine task, ordering and perfecting the earth. Spiritual reformer Francis of Assisi (1182–1226) pressed traditional theological and cultural limits with his notion that God, Man, and both animate and inanimate Nature shared a common being, yet not even Francis ever imagined an "ecosystem," much less the assemblage of ecosystems (woodland, river, marsh, etc.) into a landscape. Though Francis might value things of nature for the divine spark they shared, and while certain late medieval secular authors contrasted some pleasures found in nature (flowers, a calming view over cultivated fields, and excitement of the chase) with intrigues of the court or urban bustle, no medieval evidence values Nature for its own sake independent of human uses. Still, the actual situation of medieval Europeans in their environment no more depended on their *awareness* of ecological connections than their going to work at sunrise rested on their firm recognition that the sun circled the earth.

This chapter identifies some important interconnections between natural systems and human cultures in medieval Europe, beginning with aspects and cases where stable relations may be considered as coevolution and equilibrium. A second section then indicates how breaks in the patterns arose from natural as well as human forces, but played out in accordance with the ecological principle that all is linked to all. Medieval environmental history thus contains stories of stabilities and changes that collectively comprise one temporal slice of continual interplay between the human and the natural on the westernmost extension of Eurasia.

HUMANS AND ECOSYSTEMS IN MEDIEVAL EUROPE: CASE STUDIES IN COEVOLUTION

In northern Europe's two principal geoclimatic regions, seasonal norms of temperature and precipitation acting on distinctive topographies set dynamic stages for preindustrial human societies. The point is not environmental determinism, but rather that there were distinct and relatively predictable conditions to which successful human societies learned to adapt in various ways. During the Middle Ages people increasingly did so by simultaneously adjusting natural conditions to their own needs.

In Atlantic Europe, winter cold required native sources of edible oils other than the sensitive olive but did not force livestock indoors. The region's humid, often deep soils commonly supported dense natural woodlands and the growth of summer cereals. By contrast, in continental Europe, frozen soil restricted the success of traditional winter wheat and favored rye. Livestock had to be housed and fed or taken on long-distance migrations from summer to protected winter pasture zones (horizontal transhumance or nomadism). (See chapter 1 for additional detail on European climate during the Holocene.)

Like all farming, the traditional agricultures of medieval Europe were artificial ecosystems adapted to meet human cultural needs from available living and nonliving components of what was then the local environment. (See chapter 2 on the ecology of agroecosystems.) While south of the Alps the adaptation was largely completed during the Bronze Age, in northern Europe the low-intensity agriculture of the Roman Iron Age remained technically imitative of southwest Asian and Mediterranean antecedents. Northern European farmers both innovated and greatly expanded their activities during the Middle Ages. Insofar as their activities entailed continuing adjustments to regional conditions they are treated here; ecological consequences of creating arable are the subject of a later section.

Agricultural practices created to handle the summer heat and drought of southwestern Asia and the Mediterranean had to be modified for northern conditions. Where not moderated by the Atlantic, northern winters inhibit plant growth, but moisture available throughout the year both permits spring-sown crops and removes soluble nutrients from the upper soil layers normally accessible to shallow-rooted annual plants. During the Middle Ages in Europe, adaptations to those new circumstances for agriculture were still taking place at both the continental and the smaller, regional scales, posing different problems of sustainability and stability.

Intensification to permanent arable for cereals was an important development completed during the ninth through thirteenth centuries in most of northern Europe. Spreading use of the moldboard plow solved essential water-related problems by turning the soil over to make soluble nutrients available at the soil's surface and by enabling farmers to adjust field profiles in various ways to manage runoff and infiltration. Characteristically heavy soils and dependable year-round precipitation meant land could be cropped more frequently than elsewhere. Resting the soils only one year in three, rather than one in two, promised half again more total production at the cost of more labor to plow and harvest and a third less output of favored winter bread grains. More spring-sown barley and oats supplied more beverage and more animal feed for a technology and culture more reliant on animal traction and on meat eating than was the case in the Mediterranean. In the stereotypical "Midland system" of central England, livestock pastured on the stubble and fallow recycled some nutrients to the fields. Complex adaptations fit different limiting factors elsewhere.

Sustainability remains for traditional northern agriculture an analytical issue as well as a practical one. As elsewhere, those nutrients embodied in grain and animals which were consumed outside the farming village were lost and only marginally replaced by spreading barn litter, leaves, and herbage cut from waste ground, seaweed, marl, or urban effluents, which were taken to the fields as they occasionally became available. Even on once-virgin, rich soils for cereals, centuries of medieval use gradually lowered fertility to a plateau visible to historians in the falling yields of midland English arable under the heaviest population pressures around 1300 in locations where additional resources were unavailable. While economic historian Michael M. Postan once evoked soil exhaustion as an explanation of lowered fertility, later writers have also recognized causes such as pressures of taxation, weather events, and economic choices between fertility maintenance and other priorities. Although English records are peculiarly rich and well studied, ecological consequences of long medieval use of the soil do cry out for investigation elsewhere, especially in earlier-settled and highly productive regions such as the southern Low Countries, the Paris basin, Alsace, or the Bavarian plain. Sustainability within certain social and economic parameters—and not only in direct relation to fertility maintenance—is arguably one ground for many widely followed management practices such as plowing the fallow to keep down weeds, division of each operator's holdings into scattered strips, peasant use of available wild resources, and rationed access to common pastures (stinting).

Coevolution with regional environments was a continual process in medieval northern European agriculture, at least for people with access to more resources. Local and socially patterned innovations included planting field legumes (which, unbeknownst to medievals, fix nitrogen in a form accessible to plants) to raise fertility while feeding animals or even humans; devoting more labor to preparing and manuring the soil, sometimes from off-farm sources; and basing production decisions on marketability rather than local consumption. These techniques remained more typical of lord-run than peasant agriculture and were, whatever their effect, rarely chosen for sustainability's sake. Indeed, production that was meant to provide an exportable surplus, such as the grain shipped from landlord farms in the Thames Valley to London around 1300, or the Danish beef cattle destined for the fifteenth-century towns of northwest Germany and the Netherlands, accelerated loss of nutrients and energy. Likewise, cutting leafy fodder and spreading woodland litter or turfs on the arable impoverished nonagricultural ecosystems to subsidize the agricultural. But every such tactic also indicates incremental increases in human knowledge of how to gain and retain more from the productive relations of a locality. Progressive microadaptations to local soils, water, and climates thus shaped alternative subsistence strategies, which during the later Middle Ages challenged an earlier obsession with grain.

No medieval European agriculturalists, however, transcended existing ecological constraints on agricultural productivity. This had to await new food species of exotic origin (potato, rice, maize), intentional fodder crops (legumes, turnips, artificial meadows), systematic transfer of nutrients from elsewhere (manure, guano, synthetic chemical fertilizers), and another order of magnitude in interregional food trades. All became central human conditions for European agroecosystems in later modern times. Until then the ancient Mediterranean triad of bread, wine, and olive oil, or the medieval northern combination of cereal grains (for bread and beer) with animal products (meat, butter, and cheese) remained cultural reflections of distinctive environmental adaptations and limitations.

Different kinds of woodlands grow naturally almost everywhere in temperate Europe (alpine tundra and Pannonian and Pontic steppe excepted). Medieval Europeans used native woodland species as sources of human food, pasture, raw material, and energy, most of them well captured in an 866 charter from Worcestershire, where recipients of a wood anticipated pasturing pigs; gathering poles, fuel wood, and construction materials; and selectively logging timber trees. While pristine wilderness was always more present in the imagination than on the ground, natural plant communities, accessibility, and human population density shaped choices between multiple uses or specialized management. Long historical trends favored the latter.

Local soil and climatic conditions were the primary determinants of "original" or potential tree species: broadleaved oaks (deciduous or evergreen), hornbeam, hazel, beech, chestnut, birch; conifers such as juniper, yew, cedar, pine, fir, or spruce; and their associated vegetation communities. But historians such

Woodland in Tenth-Century Northwestern Europe

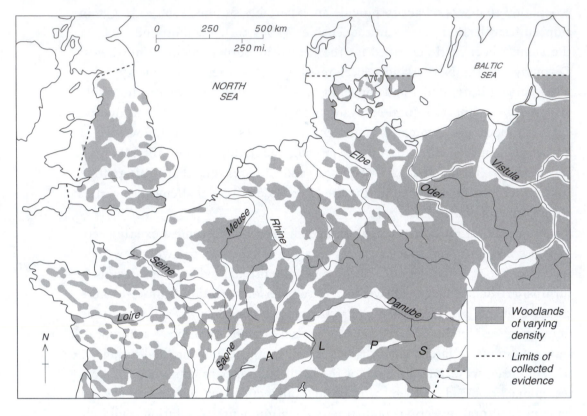

Source: Compiled under supervision of Richard Hoffmann by Carolyn King, cartographer, Cartographic Laboratory, Department of Geography, York University, using data from O. Schlüter, "Die Siedlungsräume Mitteleuropas in frühgeschichtlicher Zeit, Part I," *Forschungen zur Deutschen Landeskunde*, vol. 63 (Hamburg, 1952) and Charles Higounet, "Les forêts de l'Europe occidentale du Ve au XIe siècle," *Settimane di Studio* 13 (1965), 343–398.

as Chris Wickham have shown how the forested landscapes long recognized as widespread in the early Middle Ages actually ranged across spectra from dense to open vegetation and pristine to humanized by long cultural interaction—even at places like eighth-century Fulda, where monks tried hard to imagine a *deserta silva* ("uninhabited woods") despite evidence of older human occupation. Places such as the contemporary Odenwald or the twelfth-century *preseka* (Poland's western frontier) were little penetrated by humans. Still, by the Domesday survey of 1085, a mere 15 percent of England was wooded and, as elsewhere, this proportion fell another third or half during the following 200 years. Grounds for and effects of such clearances are considered below; the focus for now is on surviving woodland and its continuous uses, as these evolved within the limits of natural bioproductivity.

In some parts of medieval Christendom, and even past the sixteenth century in Scandinavia and elsewhere, some woodland served as a reserve of temporary plowland, cut and burned for a few years' farming in the fertile surface ash, then regenerated on a long rotation as "forest fallow" for many more years. Woodland was pasture, too, for livestock like to eat the new growth of most woodland species and the seasonably abundant fruits of oak and beech. In many regions, browse in wood pasture (*silva pastilis*) supported swine and cattle—both naturally creatures of woods and woody edge. Herders could gather "leafy hay" by lopping lower branches or even climb to cut them from taller "shredded trees," which left the trunk for timber.

The fundamental distinction in woodland use and exploitation was between timber and coppice. Timber woods held big old trees, commonly oaks or beeches, perhaps conifers in some mountain regions, which were suitable for heavy construction but which, once cut, took some 50 to 150 years to regenerate. Maturing timber trees soon grew out of the reach of animals, so the same woodland also pastured domestic beasts and provided habitat for game animals, for understory plants (berries, mushrooms, herbs, etc.), and for other organisms (bees) that humans used. Branches from felled timber fueled local hearths, but whole mature trees were too valuable and too ungainly to be large-scale sources of fuel. Up to about 1000 CE timber trees seem to have been sufficiently accessible in most of western Europe so that few surviving records specify their management, but from the thirteenth century most landowners actively protected and controlled these valuable investments. This was possible because medieval Europeans knew how to use quite different sorts of woodlands to meet their needs for energy and raw materials.

Coppice was the essential element in medieval development of managed woodlands integrated into an advanced preindustrial economy. Coppice exploits the natural regenerative capacity of most woody vegetation (but few conifers) to respond to cutting by putting forth shoots or suckers, what vernacular English calls "spring." This active growth, very palatable to animals, will in a few years, depending on the species, yield poles of thumb-to-forearm size and some meters in length, suitable for wattle, rods, stakes, poles, bundles or blocks of fuel wood, or conversion to lighter, pure-burning charcoal. While charcoal yields two-thirds more calories than an equal weight of wood, medieval producers reduced three tonnes of wood to one tonne of charcoal. A sixteen-year-old oak coppice produces eighty-four cubic meters of wood per hectare. Medieval coppice woods were, depending on species and intended uses, managed on four- to twenty-nine-year cycles, with a ten-year mean documented in France. Where browsing animals threatened the coppice, trees cut some two to four meters above ground formed a "pollard," whence poles grew safely out of browsers' reach and the woodland could serve multiple uses. Some coppice could also produce timber by selecting stems to grow as "standards" through several successive coppice cycles.

Pollard oaks can live a millennium, and other species can live up to 500 years; coppice stools (the underground portion) can live and produce shoots indefinitely. Some deciduous oak pollards still producing in Sardinia date back to late-medieval centuries, and certain ash stools in Suffolk may be the oldest living things in Britain. The management practices of "woodmanship" are even older, though since it is normally the business of illiterate workers, these practices are rarely objects of literate record. Traces of the medieval practices of woodmanship must now be winkled out of rare written references, archaeological fragments, and historical botany itself. A rare 863 charter from the Weald of Kent alludes to seasonal rules for taking firewood poles, but generally uncovering past methods of woodmanship is a major research enterprise in most regions.

Plantation forestry was less common, or at least subject to elite attention only from about the fourteenth century. An ordinance from French King Charles V in 1376 required woodcutters to leave "seed bearers" for natural regeneration. By about the same time managers of Nürnberg's city forest were selecting seeds of pine and fir for germination and eventual plantation. Plantation woods tend toward single-use artificial systems, eventually akin to agriculture.

Medieval use of woodland thus relied on detailed knowledge of environments and tree varieties. For example, eighth- through tenth-century charters from the abbey of St. Gallen and the bishopric of Freising show inhabitants of the south German montane zone choosing specific trees to take wood they needed for certain artifacts and structures; they cut leaves and drove swine to oak or beech mast for fodder; hazel coppice provided nontimber raw materials and fuel. Contrasting with these carefully managed woodlands were practices in late medieval and early modern Scandinavia—the best evidence is from Finland—where comparably selective and continual requirements for wood to serve ordinary subsistence purposes meant people placed little or no value on conserving it for future use. "Rational" legislation against depletion remained wholly ineffectual until timber acquired commodity value in the nineteenth century.

The history of medieval woodlands under continual use verifies the success of these mutual adaptations of human uses and woodland vegetative communities. The intensely conservative, medieval English managed woodlands as a renewable resource. Those woods which survived into the late thirteenth century were subsequently changed little, and that but gradually, up to the twentieth century. In France the industries with notably heavy demands for fuel were relatively more mobile, shifting from region to region as they depleted accessible supplies and became subjected to governmental regulation for the sake of general consumer demand. This is not to ignore the real effects human use had upon species composition and woodland structure. Taste preferences and the physical impact of grazing animals encouraged replacement of unprotected woodland species with thorny scrub or managed coppice. Traffic through a regularly harvested coppice compacted the surface and encouraged formation of soil horizons; its very raison d'être held the woodland ecosystem at an artificially "young" pioneer stage. Coppicing for fuel in the natural, mountainous beech-wood regions of central Europe (such as the Black Forest, Vosges, Siegerland, and Harz) slowly selected against beech and favored the regrowth of oak and birch underwood. Because young plants have higher nutrient levels than do large trees, which are mostly nonliving wood, the typical consumption of coppice products at sites elsewhere depleted the soil more rapidly than did timber. Yet even timber woodlands operated for harvest at intervals of 100–150 years contained minimal old growth and dead or dying trees, so they provided little living space for organisms with those habitat preferences. "Parks," "chases," and "forests," that is, lands subject to special legal jurisdiction for management as game preserves of nobles or kings, could retain relict tree species, but natural densities were thinned to ease physical access. Some northern hunting preserves more resembled the open wooded savannas indigenous to the Mediterranean.

It seems, however, that neither private ownership nor common access regimes led inexorably to depletion or effective conservation in medieval woodlands. Aware that their resources could sustain only limited levels of use, all across Europe peasant communities struggled to limit access, especially by new village households, to pasture and woodland commons. Indeed, the real historic "tragedy of the commons" was not destructive overuse by those who had common access, but actual expropriation by lords and private landowners who might sell or log woodlands for a quick cash return. But in numbers of other cases, private ownership did sustain limited use of renewable resources for centuries, while open public use was depleting other woodlands nearby. Based on present evidence, it appears that the combination of demand and access permitting the economic movement of wood to consumers had more influence on sustainable practice than did the institutional form of ownership rights. Case studies from Germany, Scotland, and Spain argue that access to markets and consequent loss of local control brought critical overexploitation. By contrast, intensive private coppice production in thirteenth-century Kent remained sustainable in the long run despite its close integration into the highly competitive fuel markets of London, Flanders, and Normandy.

Patterns of coevolution between humans and woodlands thus tended in various ways to balance during the Middle Ages, as humans learned to mesh their uses with the ecological productivity of natural organisms and processes. These were, to be sure, rough and contingent equilibria, especially when seen through characteristically scant historical evidence. While northern Europe underwent great *de*forestation during the tenth through thirteenth centuries to enlarge permanent arable, in surviving woodlands purposeful techniques of management continued and even spread. Already by the 1200s in the west, European woodlands were actively protected while continuing to serve a variety of essential human purposes.

ECOLOGIES OF CHANGE: NATURAL FORCES AND HUMAN ENVIRONMENTAL IMPACTS IN MEDIEVAL HISTORY

Historically demonstrable instabilities actually confirm the complex linkages between environmental and human conditions in medieval Europe. Both human activities and natural phenomena disrupted equilibrating adaptations and provoked cascades of changes and responses, themselves shaped by antecedent natural and sociocultural relations.

Certain grave occurrences over both the short and long terms had indubitably nonhuman origin, though humans and other organisms necessarily reacted. Recent findings from examination of tree growth rings around Europe and elsewhere, for example, clearly identify several periods in which vegetative growth was generally and sharply interrupted for periods of some years, even decades. Once alerted, historians can recognize in medieval records human awareness and broader consequences of such global perturbations.

For example, the growing season of 536 CE initiated five to fifteen years of remarkably narrow rings in series from several different tree species taken across Europe and elsewhere in the northern hemisphere. Contemporary witnesses in the Mediterranean and in East Asia described (and habitation sites and glacial ice cores from Peru confirm) what today's atmospheric scientists would call a "dust veil event," which sharply cut solar radiation to the earth's surface. These events result from large volcanic eruptions or impacts of extraterrestrial objects, though no particular occurrence in 535 has so far been convincingly identified. Reduced solar input causes general atmospheric cooling, disrupted weather patterns, and shrunken biological productivity, including food crops. (See chapter 5 on the effects of the explosion of Tambora in 1815.)

Records widely report storms and cold during the year 536–537 and also for later years, followed by crop failures, a great drought in Italy in 539, and the spread of a devastating epidemic across the Mediterranean and into western Europe from 542. Given the thin sources for the history of everyday life characteristic of the early Middle Ages, at some point further alleged consequences in politics and culture become an untestable chain of contingencies. Following subsistence crises in the late 530s, unprecedented numbers of Slavs crossed the Danube into the Balkans, and the Avar tribe left Mongolia on what became a generation-long migration to the west. British resistance, which had held Anglo-Saxon invaders in the eastern third of the island for some decades, quietly disappeared in the mid-sixth century. By shortly after 600 CE, European populations had dwindled to a postclassical low point.

Comparable environmental peculiarities marked the entire period from 1314 to 1353. All tree ring series for European oaks show precipitous collapse of growth rates in 1318 and unusually low levels until 1353. Human records identify peculiarly savage weather during the period from 1314 to 1317; this triggered mounting crop failures and epizootics among domestic livestock, which laid in turn the material conditions for widespread famine and mortalities from 1315 to 1319. A few well-studied areas show extraordinary storminess, marine inundations, and earthquakes in this and following decades. Other ecological oddities during the second quarter of the fourteenth century need also to be laid against evidence of very high human populations in much of western Europe. But the huge mortalities of the Black Death, 1347–1351, which precipitated Europe into a century-long demographic free fall, are not convincingly associated with prior malnutrition or high human densities. This is true whether or not the first epidemic and early recurrences were, as thought through most of the twentieth century, manifestations of plague (*Yersina pestis*, chiefly in its bubonic form but possibly with high frequency of the more virulent pneumonic

Burying victims of the Black Death at Tournai, 1349. From the Annals of Gilles de Muisit, *1352, Abbot of Saint-Martin (d. 1353). (Snark/Art Resource)*

and septicemic plague), or if they were instead some unknown disease or combination of diseases, as scholars bothered by discrepancies between modern medical descriptions of plague and eyewitness medieval reports of symptoms and incidence have recently proposed. In any case, during the first half of the fourteenth century, relations among humans, food organisms, and parasitic disease organisms were repeatedly disrupted and with immense long-term effects.

Identification and consequences of relatively short-term events in the biosphere and atmosphere around the start of the fourteenth century are complicated by what looks to be an episode of global climatic change, the transition from a "Medieval Warm Period" (about 900–1250) to the "Little Ice Age," with mean annual temperatures as much as 1.5°C lower and greater seasonal differences. It now appears as if the cooling trend propagated progressively southeastward across Europe, with the British Isles and northwestern continent feeling effects from the 1290s, the Mediterranean after 1320, and central Europe during the mid-fourteenth century. Apart from greater storminess, global cooling meant different risks in different geoclimatic regions. On the Pannonian plain, dangerous summer droughts are reduced under wetter conditions with less evaporation, while in the eastern Mediterranean the Little Ice Age raised the risk of spring droughts on Crete.

Consequences of climatic change are most easily followed at high latitudes and high altitudes. For example, Norse settlers arrived in Greenland from Iceland in the late tenth century with a subsistence strategy based on milk, meat, and fiber from cattle, sheep, and goats, which they overwintered indoors on summer-cut hay, supplemented by land-based capture of seals, seabirds, and caribou. Foreign goods were gained by exchange of walrus hides and ivory taken along the sea ice far to the north. The cooling climate after 1250 reduced the time livestock could feed on natural forage and the growth available for winter storage: animals had to be fed longer from smaller supplies of hay. Heavier sea ice shifted seasonal marine mammal concentrations away from the Norse settlements and closed access to far northern resources. Without the adaptation to fully marine hunting of their longtime Inuit neighbors, the Norse Greenlanders abandoned their more-exposed Western Settlement between 1341 and 1362, and they disappeared from their southern farms later in the fifteenth century. Their contemporaries in Iceland suffered terrible deprivations, deaths by starvation of animals and humans, but new adaptations replaced woolen homespun with dried codfish as a means of foreign exchange to support survival of the society. Less dramatic but more widespread were such comparable adjustments as the retreat of oats cultivation in the Lammermuir hills of southeastern Scotland from 300 meters altitude in the thirteenth century to below 200 meters by the fifteenth. Meanwhile Alpine glaciers, in general retreat since the 800s and continuously since about 1100, surged during the 1340s–1370s over roads, canals, and houses to new forward positions equaled but never surpassed in later phases of the Little Ice Age.

Climatic change is associated with greater storminess, which is manifest and most easily tracked as extreme local weather events. The 1362 storm flood called the "Great Drowning" (*Grosse Mandrank*) in North Friesland swept livestock, people, villages, and some 40 percent of the land (more than a hundred square kilometers) from the island of Strand, and even higher proportions from smaller nearby islands. Local disease events had similar qualities, including the 1464–1465 epidemic in central Silesia, which eyewitnesses say killed many villagers but few townspeople (because towns refused entry to the possibly contagious), and the puzzling outbreaks called "English sweats," which five times brought high mortalities in late-fifteenth- and early-sixteenth-century England, but only in 1529 also ravaged port cities in the Low Countries, Germany, and Scandinavia.

Although changes in atmospheric and microbiological environments are so far the most plainly documented natural variations in medieval human ecology, research into geophysical events (earthquakes and volcanic eruptions) and other possible phenomena has barely begun. However, some superficially natural phe-

Agricultural work traditionally associated with the month of March, as depicted by the Master of the Grimani Breviary, a Flemish codex created ca.1490–1510: A moldboard plow appears in the foreground and vineyard work and sowing in the middle distance, with a walled city as backdrop. (Giraudon/Art Resource)

nomena also had strong human preconditions or triggers. Human activities of economic or more purely cultural origin initiated independent waves of ecological effects. Agricultural colonization and drainage, hunting, and urbanization notably show human environmental impacts throughout medieval history, not just at a later industrial age.

In a great surge of agricultural clearances, medieval Europeans mounted their largest single assault on natural ecosystems. What some called "assarting," the grubbing out of trees, intentionally transformed multipurpose woodland to specialized, permanent plowland and carried with it large unconsidered and even unintended consequences.

The process originated on essentially cultural grounds during the early medieval period, as northern barbarians adopted the cereal-based diet of "civilized" Mediterranean peoples, and then from about the tenth century came to be driven mainly by population growth and human material need. In what is now France, woodland fell from 30 million hectares around 800 to 13 million by about 1300. A provision of the 1236 English royal Statute of Merton, reiterated in the 1269 Statute of Marlborough, restricted land use changes lest too much pasture and woodland be converted to arable. In western Europe the process was ordinarily piecemeal, so it is hard to generalize about. In east central Europe, however, the bulk of clearance activity was delayed into the twelfth and thirteenth centuries but then took place more abruptly and with more visible effects. Poland as a whole was 16 percent cultivated in 1000 and 30 percent by 1500; in some parts of Silesia, the cultivated area increased ten to twenty times between 1150 and 1350. A mid-fourteenth-century chronicler wrote that never before had so many new villages and towns been founded where once only trees had grown. All this had ecological aspects and widening consequences.

The medieval Europeans who formed new arable fields were creating and expanding a new ecological niche, where humans could capture a larger share of primary biological production. The food-generating component of human culture took energy out of the natural system. What from an economic perspective was intensification, an increase in the labor annually applied to a given land area, is in ecological terms an expenditure of energy to capture more energy. This was accomplished by replacing old natural systems, characterized by low productivity relative to high biomass of long-lived organisms, with an agro-ecosystem containing a lower biomass of short-lived pioneer plants with higher relative productivity. High diversity of producing and consuming organisms gave way to low diversity, ideally a monoculture of annual cereal grasses (with competitive "weeds") and a short food chain, namely some domestic herbivores and many, mainly plant-eating, humans (with competitive "vermin"). Ecosystems with a low diversity of short-lived organisms and truncated food webs are characteristically unstable, requiring continued inputs of energy to keep them at the pioneer stage. (This was why traditional agriculture, as described above (see pages 30–31), tilled the soil, plowed the fallow, weeded, and fertilized.) Rising human needs made medieval colonization a recurrent process, moving forward on what might be called frontiers, both local and regional, or alternately referred to as the economic margin, where returns (energy) gained could be expected to exceed costs (energy) expended.

An enlarged niche for humans changed the environment for all other organisms. Physical removal of woody vegetation transformed terrestrial habitats, altering carrying capacity for wild and domestic animals alike. From the eleventh century the share in western European food waste of swine, most dependent on woodland forage, generally declined, and that of cattle likewise, if less so. Sheep and goats, better suited for living in the interstices and edges of arable landscapes, became for some centuries the chief meat animals of the west. Deforestation affected hydrological regimes as well. A late-thirteenth-century Alsatian writer even drew the connection between loss of wooded cover and more frequent and dangerous flooding, with consequent degradation of aquatic habitats. Biodiversity plunged and ecosystems were subjected to revolutionary simplification and instability.

Meanwhile the new medieval agroecosystem altered the balance of relationships among plant cover, soil, and nutrients. Colonization and intensification initiated and accelerated large-scale soil erosion and deposition, even to the point of forming new physical landscapes at some distance from the sites of human intervention. Broken vegetative cover coupled with soil and nutrient loss due to energy exports exposed large expanses of soil surface for seasonal removal, more often by water than by wind, and subsequent deposition. In northern European alluvial zones and estuaries, the unintended changes induced by human activities were abrupt and dramatic. Rates of alluviation in the upper Thames Valley during the eleventh through thirteenth centuries surpassed all other postglacial periods. Bottom cores from Lac d'Annecy (Savoy) show a sharp jump in sedimentation at the precise point in the thirteenth century when local monastic estates converted from woodland to arable exploitation. As large-scale clearances at the top of the Vistula watershed changed its hydrographic balance during the thirteenth century, at the watershed's bottom the silt and nutrients washed from new inland fields filled the historic bay between Gdańsk and Elbląg, created a chain of barrier islands, and arguably contributed to commercial failure of the herring schools that had for centuries spawned along the Pomeranian shore.

Medieval drainage projects may have had more dramatic effect than "deforestation," though in more limited areas. Cultural aversion to wetlands reinforced local pressures for more pasture and arable. At first simple embankment and ditching dried out coastal and estuarine Low Countries and interior zones of interrupted drainage in Germany, England, and France to support agricultural settlement and production. Early experiences of Flemings and Hollanders made them sought-after colonists across the North Sea and in central Europe. Conflicts between local inhabitants and incomers reveal destruction of common wetland resources, fisheries, fowling, and marsh vegetation.

Long-run consequences were especially vivid in proverbially reclaimed Holland, where gentle domes of waterlogged peat and bog vegetation had, since the retreat of Pleistocene glaciers, grown up between coastal dunes and sandbank-lined rivers. During the early Middle Ages this landscape supported a Frisian economy of wild resources, pastoralism, and waterborne communications. From the late ninth and tenth centuries, acculturation to a Frankish norm of cereal eating encouraged the practice of using simple ditches to let the water drain from the bogs, permitting first cattle and then plow agriculture on the deep, nutrient-rich peat soils. In time local communities organized joint construction and management of larger, more complex drainage schemes. When peat dries, the organic matter oxidizes and the soil layer contracts at a rate of a centimeter per year, a meter per century. By the late thirteenth and fourteenth centuries, Dutch landscapes that had once stood some meters over sea level were subsiding below it. In other localities mining of peat for fuel further sank surfaces beneath the water table. Especially during storms and high tides, the land now required protection by seawalls and dikes. Then rain or river water that accumulated behind the dikes had to be pumped up and out. Wind-powered pumping engines became typical of Holland during the fourteenth century as a direct consequence of environmental impacts resulting from human actions undertaken several hundred years before. The anthropogenic hydraulic regime further established conditions for the development and spread of endemic malaria along the North Sea coast.

Nearly everywhere in later medieval Europe, ecosystems and landscapes showed indelible marks from a great, half-millenium-long expansion of human intervention for the sake of cereal foods. While irrigation works had little importance in naturally well-watered medieval northern Europe, manipulation of rivers for water power did affect local landscapes and ecosystems during the Middle Ages. Together with cereal farming, water mills to grind the grain proliferated across the high medieval north. Especially on higher-order (that is, small tributary) watercourses, the unreliable run of the river flows encouraged storage behind dams where the falling water could drive an overshot wheel as human needs dictated. Mill dams thus shifted seasonal flows, turned running into still waters, encouraged warming and siltation in the basin, and—as Scottish lawmakers who required a passageway for salmon were well aware—impeded migration of aquatic organisms.

Hunting of wild animals provided material needs (meat, hides, bone, antler) and culturally valued leisure to medieval Europeans. Yet medieval hunting practices were culturally complex. For many elite males, hunting showed off their high social rank and maintained boundaries against their inferiors. On the other hand, various local or regional norms also allowed ordinary people to participate in some ways. Exceptions occur often enough that the stereotype that only elites could and did hunt in medieval Europe should always be tested on firmly dated regional evidence, which may, indeed, reveal ambiguities. Early medieval written records from northern Italy and northern France (Lombardy and Francia) make hunting seem socially indiscriminate, but whereas French peasant village sites from this period do show significant use of red deer (*Cervus elaphus*), contemporary and comparable Italian settlements do not. By the twelfth through fifteenth centuries, remains of game animals are a significant component in northern French elite sites—up to 69 percent of mammal remains at Chevreuse Castle—and all but absent from urban and peasant food remains. English finds of like date replicate the French pattern. But only in the late fourteenth century were western European commoners barred by law from hunting game even on lands outside designated elite game preserves (forests, parks, chases). And quite apart from game, some animals—crows, ravens, foxes—were actively pursued as vermin.

Hunting pressure caused local and regional extirpations. Large carnivores—bears and wolves—were driven into the larger mountain ranges of continental western Europe (such as the Pyrenees, Alps, and Apennines), and other impressive species were hunted down as trophies. In 1627 the last known auroch was sought out and killed as a specimen in Poland, where a relict population of the European bison also managed to survive. Among furbearers, the native beaver was so persecuted in western and southwestern Europe as to become wholly unfamiliar to most late-medieval naturalists. The marten, which twelfth-century texts treat as an expected presence in wooded regions from Britain to the Balkans, had by the fifteenth retreated to the eastern Baltic, Balkan Mountains, and Russia.

Overfishing also affected what had been favored freshwater and migratory fishes. Between roughly 700 and 1200 along the south coast of the Baltic, sturgeon (*Acipenser sturio*) fell from 70 to only 10 percent of the fishes consumed, and the mean size of specimens diminished as well. To the west, written and archaeological records of sturgeon decline steadily from the eleventh to the early fourteenth centuries. No Dutch remains of sturgeon postdate 1000, and in more than fifty French excavations they occur only at two royal sites. Indeed in

The hunt as a cultural signifier, from a mid-fifteenth century German woodcut. (Hulton Archive/Getty Images)

France and England alike at this time, sturgeon gained the rarefied status of "royal fish." The salmon (*Salmo salar*), which were given out freely in the eleventh century by lords of rivers in lower Normandy, were controlled with parsimonious care by about 1300; nevertheless, by 1423 the catch on the best-documented stream had dropped by a further two-thirds. A London record from 1386 said simply, "Salmon and sturgeon are completely destroyed."

One response to depletion was to regulate fishing with the intent of conservation. By the mid-1200s, French kings and English county courts were controlling seasons, gear, and minimum size of capture. Another response was to turn more attention to seasonally abundant marine fishes, beginning the trend to larger, more heavily capitalized commercial fisheries in ever more distant waters that would extend globally up to the late twentieth century.

Wild animals were preserved when and where elite social interests were engaged. Since Carolingian times monarchs had established their own "forests."

Foresta designated a royal hunting zone under special jurisdiction to protect and manage selected species—notably deer—together with their habitat. Fallow deer (*Dama dama*) occur almost exclusively at royal sites in France, where kings ordered local peasant tenants to harvest and deliver hay to feed them and—as at Fontenay—to build spring-fed pools for their water supply. Great barons imitated royal initiatives with their own parks and warrens, and private hunting reserves were gradually also emulated by lesser nobles. By 1500, English gentry (who would rank as lesser nobles in other societies) had created more than fifteen private deer parks in the county of Cornwall alone. Venison was a recognized luxury food, served at feasts and to honor special guests. King John of England also gained good incomes by having his foresters round up and sell his deer for their meat. Some scholars now even argue that the regulatory regimes came to be enforced as much for profit from fines and licenses as for any conservation purposes.

Medieval elites also introduced exotic animals to European landscapes. Anglo-Norman aristocrats brought fallow deer and pheasants to England in the early 1100s. The rabbit (*Oryctolagus cuniculus*), native to North Africa and brought to Spain by the Romans, was by the twelfth century common there and in Italy and southern France, while landowners in England and northern Germany sought stock for their own estates. By the end of the Middle Ages rabbits had crossed the Vistula and penetrated the Hungarian plain. The common carp (*Cyprinus carpio*) slowly expanded from an ancient native range in the Balkans into the upper Danube and Rhine basins during the tenth through twelfth centuries and then more quickly into west-flowing watersheds of northern France around 1200. At least the latter two species spread in complex cultural contexts of food taboos, multiple uses, and modification of landscapes for rabbit warrens and fish ponds. Alert examination of material remains (bones, scales, and structures) as well as written record sources (account books, taxes, craft regulations) corrects popular myths and traces both planned and quite accidental diffusions. Rabbits and carp had strong ecological impacts in western Europe, both intended—as carefully husbanded substitutes for depleted wild sources of human food and furs—and as unexpected results of feral populations interacting with natural systems already under stress from human agricultural development. In a survey of sixty-four French sites with remains of these animals, the rising share of rabbits—from none in the ninth century to 90 percent in the fifteenth—precisely reciprocates the disappearance of hare (*Lepus capensis*), which occurs in all early sites and only 5 percent of the latest sites.

Northern Europe's first great wave of urbanization occurred in the Middle Ages. Up to the eleventh century, northern European society was overwhelmingly rural, with nonagricultural people limited to tiny traditional centers of church administration (bishoprics, many at derelict Roman sites, and monasteries), military strongholds of princes and warlords, or occasional, often-ephemeral trading posts where land or water routes linked different cultural groups. Beginning around 1000 CE and in a setting of improving order and rising human numbers, some such forts, territorial rulers, crafts, or trading opportunities variously drew people to more permanent nucleations. This growth was haphazard, slow, and halting for generations up to about 1200, when only three northern centers, Paris, London, and Köln, held more than 10,000 inhabitants. Then in the thirteenth century cities exploded almost everywhere: scholars now accept that around 1300, Paris numbered 200,000, London nearly 100,000, Ghent 56,000, Köln more than 40,000, and Bruges about 30,000, while dozens of places exceeded 10,000 and scores of little towns peppered the western landscape, though thinning out to the east and north. Urban centers had become the essential nodes of an exchange economy, staging points for long-distance commerce and markets supplying goods and services to surrounding agriculturalists. Even military elites with nonurban power bases spent increasing time (and money) in towns. Modern Europe's basic urban network was largely established during the high Middle Ages, and its medieval foundations were greatly modified only in the industrial age.

Urban populations suffered in the demographic and economic crisis of the fourteenth century—London shrank to 40,000 and Paris below 100,000 by the early 1400s—but rebounded smartly thereafter. By about 1500 a half-million townspeople comprised 40 percent of the population in Flanders, Brabant, and Holland, northern Europe's most urbanized region. Once secondary centers in central Europe also attained higher rank: Nürnberg, only 14,000 in about 1300, held 36,000 two centuries later.

Urban living created a new human environment in medieval northern Europe, one very largely built and densely inhabited, where most people engaged in nonagricultural occupations and continually exchanged materials and energy with a differentiated "hinterland" outside the walls. Cities crowded humans and commensal organisms together in physical surroundings purposely modified for human social needs at an unprecedented scale. Vernacular rural wood, wattle and daub, thatch construction, unpaved passageways, and stored foods and raw materials of natural origin provided ample habitat for rats, mice, sparrows, and certain exo- and endoparasites, as well as dogs, cats, pigs, and horses, mitigated only by frequently recurring fires. Close proximity encouraged easy transfer of parasites between households and host species. Urban authorities, whether traditional lords' officers or self-governing communes, were well aware of the hazards. They promoted planning, sanitation measures, and exile of offensive and dangerous trades to downstream or extramural locations, but paved streets and fire-resistant stone, brick, and tile still came first to the best commercial, administrative, and residential quarters. The city center remained more prestigious and salubrious than laborers' dwellings crammed beneath the walls or clustered around city gates. Nevertheless, and not just from inhabitants not knowing how some diseases were communicated, medieval cities always had distinctly higher death rates than did the countryside. Established mercantile and craft families could endure for long-lived generations, but a large floating population of impoverished newcomers turned over rapidly, soaking up rural reproductive surpluses. Medieval townsfolk also remained in regular contact with the agricultural life of the vicinity, and not just because most cities were full of gardens, vineyards, and domestic livestock.

More powerfully than peasant communities and at vastly greater scale than rural elite households, medieval towns extended human environmental impact beyond their own immediate surroundings. Centers of market exchange transmitted demand pressure to natural local ecosystems at some remove from the urban consumers, so initiating on a modest scale effects now attributed to globalization. City populations needed foodstuffs from outside, sometimes far outside, their walls. Around 1300, for example, Londoners' annual demand for 34,000 to 35,000 metric tons of grain for food, drink (ale), and animal feed absorbed what was grown on 10 to 15 percent of the productive land in a zone extending some 250 kilometers along the Thames and its estuary and reaching about 30 kilometers inland from cheap water transportation. London prices were there the norm and the main factor deciding agricultural production. At the same time the Flemish towns, collectively consuming five times more grain than London, not only pulled surpluses from neighboring Artois, Picardy, and Hainault, but also large shipments traveling hundreds of kilometers via Hamburg from newly cleared farmland along the middle Elbe. Falling urban populations after 1350 put a stop to this trade, and much arable in western Brandenburg was abandoned. On the other hand, greater postplague wealth per capita sparked a great rise in discretionary consumption of meat. Concentrated urban markets were in the 1400s fed not only on local livestock but by annual overland drives of thousands of head from Hungary to southern Germany (and Venice). Comparable Polish and Scandinavian cattle exports to the Rhineland and Low Countries followed. Staple export markets for live cattle changed late medieval land-use patterns in Hungary and Denmark. Already by the early 1200s concentrated luxury demand for fish in Paris, northern Europe's greatest market, had notably so depleted local fresh waters that elaborate, fast packhorse trains were set up to bring marine catches nearly 200 kilometers from coastal Normandy in still (barely) edible condition. A hundred years later, brined herring from the North Sea and dried cod from arctic Norway or Iceland also be-

The fair of Lendit held in June at St. Denis outside the northern gate of Paris has drawn merchants from France and other countries since before the eleventh century. In an early-fourteenth-century miniature, the Bishop of Paris gives his traditional blessing as merchants and townspeople go about their business. (Snark/Art Resource)

came staples on the Parisian market, thousands of kilometers from where the fish were captured. Only townspeople so regularly and massively exploited ecosystems so far beyond their own view.

Concentrated medieval urban demand for heat energy organized and sustained intense coppice woodland management on thousands of hectares close to London and Paris—keeping transport costs down—and the first commercial exploitation of fossil fuels. Londoners burnt "sea coal" brought by boat from Newcastle around 1300 and again after 1550. Big towns in Flanders and Brabant were already by 1200 the principal consumers of dried peat mined from deposits in North Flanders, North Brabant, and Holland some 20 to 100 kilometers away. On most soils removal of peat hampered future agricultural use of the land, turning marshy or boggy landscapes into sandy heaths or open-water lakes. Anthropogenic small lakes remain distinctive landscape features and ecosystems in the peat lands of North Flanders; South Holland's vast but equally artificial Haarlemmermeer, located between Haarlem and Leiden, not really drained until the nineteenth century, also resulted from late-medieval peat exports for urban energy needs.

Urban densities concentrated human, animal, and production wastes to pose another environmental problem. Medieval town governments not only sought inputs from outside, they also tried to manage sanitation within the walls and purposely directed effluents out into their surroundings. Rouen and Nürnberg placed latrines directly over the watercourses flowing through town. Municipal regulations at Köln required the contents of cesspits to be emptied into the Rhine through downstream gates kept open overnight for that purpose. Suburban gardeners and vineyard operators, however, might also eagerly recycle the nutrients in urban "night soil." In fact, medieval material culture recycled and reused many objects—building materials, clothing, animal bone—and relied primarily on materials of natural organic origin, so a biodegradable waste stream held comparatively less toxic material than it did in an industrial age.

Still, medieval townspeople complained about pollution—Paris colleges objected to butchers who dumped slaughterhouse waste and Londoners complained about coal smoke, for example—and some strong, objective evidence of local environmental damage survives. Archaeological excavation of the onetime Bodensee shore at Konstanz revealed clear water algae of the thirteenth century replaced by species indicative of excess nutrients by the fifteenth. A monastery beside Douai sued clothier Barru Lourdel in 1452 after his processing of flax and hemp killed fish in the local brook.

While medieval Europeans did work toward what we might call sustainable human niches in European ecosystems, changing human impacts and natural forces combined meant there was likely no pristine equilibrium or unaffected Nature in medieval Europe, just different degrees of coevolution.

4

DENYING MALTHUS
Demographic, Economic, and Environmental Developments from 1500 to 1800

In New Jersey, the proportion of births to deaths on an average of seven years, ending in 1743, were as 300 to 100. In France and England, taking the highest proportion, it is as 117 to 100. Great and astonishing as this difference is, we ought not to be so wonder-struck at it as to attribute it to the miraculous interposition of heaven. The causes of it are not remote, latent and mysterious; but near us, round about us, and open to the investigation of every inquiring mind.

THOMAS MALTHUS, *AN ESSAY ON THE PRINCIPLE OF POPULATION* (LONDON, 1798)

At the end of the eighteenth century, when Thomas Malthus wrote his famous essay, he was trying to find "natural laws," like the law of gravity, that could explain the perpetuation of poverty in the world through the idea that population was growing in a geometrical ratio while the means of subsistence grew in an arithmetical ratio: "I say, that the power of population is indefinitely greater than the power in the earth to produce subsistence for man." Because of that, subsistence crises loomed large in his essay. According to his numbers, births exceeded burials in the proportion of 124 to 100 in Elizabethan times but only 111 to 100 in his own time. This could be explained by what he called an "accommodation" to average subsistence levels in his country. In the United States, however, land was so abundant that birth rates rose high above death rates. This state of affairs could not, in his view, last forever, for sooner or later the American economy would also mature.

For Malthus, subsistence crises were a hard reality. His essay contains numerous references to hunger, epidemics, war, and "sick years." These were what he called the "positive checks" upon the growth of population. He did not

Engraved portrait of Thomas Robert Malthus (1766–1834), English economist and author of "Essay on the Principle of Population" (1798) and "Principles of Political Economy" (1820). (Hulton Archive/Getty Images)

believe that the coming of the manufacturing age had bettered the situation; on the contrary, he added "vicious customs with respect to women, great cities, unwholesome manufactures" and "luxury" to the list of positive checks because they would only increase mortality.

Malthus's world was therefore a world of limits in which the transition to a more modern economy, less dependent on land and more oriented toward industry and cities, was not a way out. His observation reflected the fact that without the importation of grain from abroad, his own country would have hovered on the brink of a Malthusian crisis: in the eighteenth century the English population grew faster than grain production. Moreover, death rates in early modern cities were extremely high. "Cities are the graveyards of the human race," wrote Jean-Jacques Rousseau (cited in Clout 1977, 489).

Despite the pessimism of Malthus, northern Europeans at the end of the early modern period were successfully struggling to surpass the limits set by an already hard-pressed environment. Subsistence crises and hunger—although still possible, as proven by the harvest failure in France in 1789—were on the retreat. Life expectancy, our best historical indicator of health and material progress, was slowly improving. Epidemics became less devastating, as did wars. Europeans were turning to new resources in order to overcome old shortages, and they were developing North America as a terrain for European immigrants. This chapter will describe environmental changes, sketch the problems that confronted the growing population, and outline the directions taken toward their resolution.

DEMOGRAPHY

Taking the long view, population in northern Europe grew. In England, the Netherlands, France, and Germany, human numbers almost doubled between 1500 and 1750 from 28 million to 52 million. In the second half of the eighteenth century growth continued, reaching 60 million around 1800. As far as we know, life expectancy rose modestly. Among French and English women it rose, respectively, from 26 and 32 to 32 and 37 between 1740 and 1790. A similar upward curve marked the nineteenth century.

More people lived in cities at the end of the early modern era than at the end of the Middle Ages. The most remarkable area in this respect was Holland, where urbanization rose from an already high 44 percent in 1525 to 60 percent around 1800. The two biggest cities were Paris and London, each with more than 500,000 people around 1750, having grown from a combined population of about 150,000 at the end of the Middle Ages. Only a few new towns

Paris as it looked in the seventeenth century. In The Architecuture of Paris XVII *by Henri Lemonnier, seventeenth century. (Library of Congress)*

had since been founded; urbanization largely took place in and around the existing urban cores.

Population growth was not an even development, however, and it stagnated in most European countries in the seventeenth century. Due to the devastating Thirty Years War (1618–1648), the German population even fell back to its late-fifteenth-century level. Nor was urbanization a linear process. The largest cities—London, Paris, and Amsterdam—grew rapidly between 1500 and 1650. London and Amsterdam grew tenfold, from 40,000 and 14,000 to 400,000 and 175,000, respectively, while Paris underwent a fourfold increase from 100,000 to 430,000 people. In all three cities, growth slowed considerably between 1650 and 1750.

By the second half of the eighteenth century, population growth in Germany and England had resumed thanks to high birth rates. In England and the southern Netherlands, later Belgium, the population had grown by about 1 percent per year since 1740. In France, by contrast, birth rates dropped, and the population grew more slowly in the eighteenth century than in neighboring countries. Scandinavia remained a sparsely populated region. Sweden, for instance, containing close to twice the area of the United Kingdom, had only 1.8 million inhabitants in 1750. With the exception of Iceland, burdened by natural disasters and famine, the Scandinavian population did grow in the seventeenth and eighteenth centuries, particularly after 1721, the end of the Great Northern War (1700–1721). Denmark's population increased rapidly, from a half-million around 1650 to more than 925,000 around 1800. Finland's population skyrocketed. It almost doubled from 420,000 around 1750 to 833,000 around 1800. Norway experienced slow growth before 1750 and more rapid growth after 1750, spurred by declining death rates and increasing birth rates. Urbanized Holland, however, proved to be an unexpected exception to the rule. In the Dutch Republic the population swelled until 1650, but then it stagnated utterly through the end of the eighteenth century.

Since most northern Europeans lived from the land, it is clear that fluctuations in human numbers had consequences for the environment and resource use. Population growth and urbanization in the sixteenth and eighteenth centuries placed the environment under stress, while the seventeenth-century slowdown somewhat relaxed the pressure.

AN ENVIRONMENT UNDER STRESS

The Little Ice Age

One of the problems that confronted the European population was the Little Ice Age. Recently historians have tried to pinpoint the consequences of lower temperatures. How vulnerable to changes in weather patterns was the largely agrarian economy of early modern Europe? That there was a Little Ice Age is now beyond discussion, but the consequences are not entirely understood. Generally speaking, the effect of climate change is that it alters the costs and benefits of a particular climate to which a region has adapted itself. English wine growers, for instance, had to adopt new crops when the wine border moved south in late medieval times. People in the Alpine regions faced advancing glaciers that destroyed crops and buildings. Norse settlers abandoned their foothold in Greenland because of the shortening of the growing season. These are rather obvious effects of the lowering of temperatures, yet the Little Ice Age was not a smooth and gradual affair. The process was more fickle. Temperatures in the northern hemisphere dropped significantly, by approximately one-half degree Celsius after 1400 with respect to the mean during the period 1900–1960 (before greenhouse gasses markedly influenced climate). By the late sixteenth century, however, harsh winters and cool summers in central Europe brought average temperatures of fully two degrees Celsius lower than the 1900–1960 mean temperature. The 1690s and the first decade of the nineteenth century were also very cold. Summers in the 1700s were, on the other hand, warmer and wetter than those in the first half of the twentieth century. Another important phenomenon was that the Little Ice Age was accompanied by more extreme weather events, such as severe droughts, pouring rains, and stronger storms.

The most erratic cause of lower mean temperatures over the long term was several volcanic eruptions, which hurled large amounts of dust and sulfur into the atmosphere, blocking the sunlight. Extremely cool summers were the inevitable short-term consequence of the explosion of Huaynaputina in Peru in 1600, a fascinating but ominous event. The sun turned red in Europe, and summer sunlight was so dim in Iceland that it produced no shadows. Eight such severe volcanic eruptions followed by cold summers were registered in the seventeenth century. The famous Maunder minimum, the disappearance of sunspots in the period between 1645 and 1715, also coincides strongly with the lowering of mean temperature in the northern hemisphere. A third phenomenon strongly affecting European weather patterns was the behavior of the so-called North Atlantic Oscillation (NAO), a kind of Atlantic El Niño. If a high prevails around the Azores and a low around Iceland, warm Atlantic winds create mild winters and stronger westerly storms in northwest Europe. When the situation is reversed, and a high prevails around Iceland and Greenland, cold arctic winds can easily penetrate the continent and the British Isles. The latter occurred more often in the late seventeenth century, creating extremely cold winters and dry summers such as the one of 1666, which abetted the Great Fire of London.

Climatologists and historians have ingeniously used thousands of weather observations to investigate the overall effects of the Little Ice Age and their strong regional variations. It is a rather complicated affair. For example, harvest failures are caused not only by cold summers but also by a lack or an excess of precipitation (not to mention marauding armies). Some crops, such as buckwheat, which was gaining ground in northern Europe, are extremely vulnerable to bad weather, especially late spring frosts or hailstorms, while other crops are more resilient. How daily life was influenced by lower temperatures is also not a simple matter. When extreme weather conditions coincided with other adverse circumstances, the effects were catastrophic: in 1696–1697, a combination of hunger, diseases, and government indifference wiped out one-third of the population in Finland. Strong gales ripping into eroded lands and dunes caused sandstorms in England and Denmark in the late seventeenth century, turning arable land into small deserts. Harsh winters sometimes influenced military campaigns, for heavily armed troops could easily cross frozen rivers, as the

French armies proved in 1672 and 1794 in Holland. Ships could not navigate on frozen rivers and canals, so longer winters were drawbacks for tradesmen, a disadvantage not mitigated by easier skating or the use of sledges. More storms meant riskier trade routes, higher insurance rates, and a temporary paradise for beachcombers.

A balanced view of how the Little Ice Age affected northern European history comes to the following conclusion. Regions with diversified agriculture and good access to the international trade network could cope comparatively easily with severe weather conditions in the form of droughts, floods, storms, or low temperatures. Holland, England, and northern France could in most cases of harvest failures import food from abroad: grain from eastern Europe, fish from the seas, and other foodstuffs from overseas, though these sources, too, were subject to the vagaries of weather. With more crops in its repertoire, the relatively advanced agriculture of England, northern France, and the Netherlands helped spread the risk of harvest failure. In the isolated country of Switzerland, on the contrary, unfavorable weather, especially cold springs and harvest rains, was strongly related to food and grain prices. In central Europe, the Little Ice Age was no laughing matter. Throughout the high Alps, with their short growing season, people felt the impact of lower temperatures severely. So did Scandinavians, who were sometimes forced to eat bark after harvest failures.

Forestry and Deforestation

The Little Ice Age was a climatic event for which no one could be blamed. However, it showed that the European economy was vulnerable. Although in the long run western Europe boasted the world's highest life expectancy, it did not come about without serious setbacks. At least one central aspect of environmental stress was self-inflicted.

Deforestation figures as one of the follies in the history of northern Europe. To a fair extent Western civilization was, literally, built on wood. The most important machines of the era—mills that were scattered across the landscape—were largely made of wood, as were houses and ships. Building a warship in the seventeenth century required 3,500 trees aged 80 to 120 years old, roughly twenty-four hectares of timberland. As brick houses became more popular, even more wood was required to bake the bricks. Although England was turning to coal and the Netherlands to peat, the most important fuel remained wood for northern Europe as a whole. Despite this obvious dependency on wood, Europeans cleared forests to create arable land, placing forests under permanent pressure.

Overview of old peat-workings in the Netherlands. (Peter Johnson/Corbis)

Numbers are very illuminating in this respect. In France forests covered 18 million hectares in 1550 and 9 million in 1789. As population increased in the same period from 14 to 25 million, less wood became available per capita. In Denmark, 20 to 25 percent of the country was forested in 1600, and only 8 to 10 percent was forested by the middle of the eighteenth century. In most regions the gradual process of deforestation had begun in the Middle Ages. The Netherlands and Scotland were already largely deforested by 1500. The wooded area of Ireland dwindled from covering about one-eighth of the country in 1600 to only 2 percent in 1700. Large-scale deforestation may even have caused climate change. Around 1800, a new phenomenon appeared: the European monsoon. Torrential rainfall succeeded long periods of drought. Monsoons occurred more often in deforested areas and disappeared in the nineteenth century when forests began to recover.

Deforestation is the final result of a process in which areas lose their tree cover. It can be a slow process or the result of quick clear-cutting. The result may be an extension of agricultural land or wasteland where forest recovery is prevented by, for example, grazing. Behind the numbers and maps lie different stories about the way deforestation took place.

In forest management, a rough distinction exists between high forest (producing timber) and low forest, particularly coppice. The contrasting nature and uses of timber and coppice have been extensively treated in chapter 3. During the early modern centuries, timber remained an industry involving large interests, namely states avid to create and maintain navies. Hauling timber over land also remained a costly affair; thus, shipyards and the building industry preferred timber that could be moved along rivers or by sea. The English navy in early modern times depended increasingly on Scandinavian timber and, for strategic reasons, on timber from the North American colonies, but this does not mean that in England timber had become scarce in an absolute, ecological sense. Rather, the costs of hauling English timber had become prohibitive. In Holland, however, where easy transport facilities were almost everywhere, timber had nearly disappeared and had to be imported from Scandinavia and Germany. The transportation of timber along the Rhine to Dutch markets reveals a fascinating picture of the scale of deforestation in the hinterland of western Europe's longest river. Trunks for the production of planks and masts were formed into massive rafts, with a maximum length of about 300 meters, a width of 50 meters, and a draught of 2 meters. Several hinged parts made steering possible on the winding river. The rafts were manned by hundreds of people, who would seek further employment on ships or elsewhere once they reached the borders of the North Sea. The pace of deforestation in Germany has been calculated as fifty times the period needed for a forest of saplings to grow to maturity. In such extreme cases, "forest management" was virtually absent. It is no accident, though, that both forest management and silviculture were first theorized in the German states.

Cutting devastated low forests less often, and many a "forest" at the end of the early modern era was probably no more than an extended coppice field, interlaced with heath or reed land. Nevertheless, this humanized landscape has often been confused with "nature": humans were usually absent, the areas uninhabited, and much spontaneous undergrowth and wildlife tolerated.

Northern Europeans relied on fuel wood to warm their houses, cook, and produce tar for industries such as brewing, brick making, and sugar refining—in short, all possible processes where heating was involved. Peat provided a real alternative only in a few regions, especially Holland (see the case study "Fuel Resources and Wastelands in the Netherlands around 1800" on pages 165–175) and Ireland, thanks to the activities of the British colonial rulers, who cleared Ireland of most of its forests. Another danger for the woodlands was the tanning industry, which required bark.

The most important quality of low forest was that it could be transformed into charcoal. It was impossible, before the age of coal, to melt iron without charcoal. Charcoal was also used in the refining of other metals, like copper, and in glassmaking. In the famous French film about the flight of Louis XVI during the French Revolution, *La Nuit de Varennes* (1982), a traveling party comes across a couple of charcoal burners right in the middle of a forest. Although the party was historically improbable (with Casanova, Restif de la Bretonne, and Thomas Paine traveling together), the presence of the charcoal burners was not. They were everywhere, especially in France, where iron industries were scattered all over the kingdom. Such decentralization sprang from the impossibility of transporting brittle charcoal over long distances and the overland transport costs involved, prior to the era of canals and trains.

In every region where iron ore was found, an iron industry could be established, providing that there was wood nearby or that wood could be grown. Ubiquitous iron allowed Sweden to become a great and feared military power in the early modern period. Although estate-owning nobles were involved in iron-making, industrialists did not necessarily need to own such estates for growing wood. Many farmers and rural artisans maintained some coppice that could be sold to charcoal burners, illustrated by just one example of a small, traditional copper mill in the eastern Netherlands. Between 1815 and 1845, 20,000 to 30,000 kilos of copperplate were produced each year. In the early nineteenth century, the entrepreneur obtained some 650 large baskets of 350 logs each per year, that is, more than 225,000 logs for one mill only, from ten to twenty different suppliers—including millers, peasants, the local baker, and tailors from the nearby village.

Because wood became regionally scarce in early modern Europe, strong incentives existed to conserve fuel. Developments in charcoal production enhanced yields from twenty-six cubic meters of wood per ton of charcoal to eight cubic meters of wood per ton. Nevertheless, problems arose, as shown by the *cahiers de doléances*—lists of grievances drafted by the three estates and other entities throughout France at the beginning of the Revolution. They abound in complaints about the ruthless exploitation of forests by charcoal burners. In the Netherlands, charcoal burners even got into conflict with peasants over the use of the moors, because they needed sod to cover the large *meilers*—cones of stacked wood taller than a person—where the charcoal was produced in a process of dry distillation. In one respect the transition to coal really lowered the

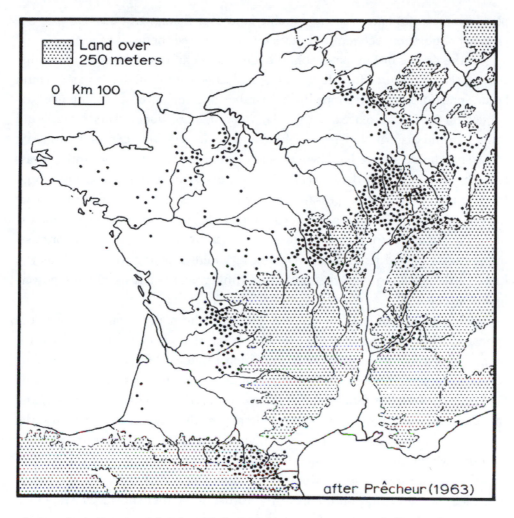

Ironworks in France at the time of the Revolution were scattered all over the country. (Hugh Clout, ed., Themes in the Historical Geography of France, *1977)*

pressure on the land, yet where forests were carefully managed to assure the supply of wood as a source of energy, substituting coal could stimulate deforestation. Land that was previously maintained as coppice now could be converted to arable land to feed a growing population.

The history of deforestation in northern Europe is not a simple story of environmental degradation, nor is it a tale recalling the fate of Easter Islanders, who undercut the very basis of their existence. Land could be used in a variety of ways: for cultivation, grazing, hunting, digging peat, growing timber or coppice, or mining. As technology developed and lakes and estuaries were drained, people sometimes opted for more arable land over fishing grounds. Despite the growing population, economic choices often presented themselves.

In the course of Europe's economic development, once-abundant wood became a scarce resource for which a price had to be paid. Whether an area would be de- or afforested, or stripped of its timber and replanted with coppice, depended on the marketplace and the relative political power of users, buyers, and sellers. In the eastern Netherlands, for example, forests were common property. Most shareholders were landowners, such as nobles or cities, and unlike peasants these landowners theoretically had a long-term interest in keeping high forest intact. Nonetheless, peasants slowly but unceasingly degraded the forests into moors. That the Dutch could easily and cheaply import timber helps explain the slack control over the forests exercised by those in power. A contrasting scenario played itself out in Denmark, where a strong party, such as the king, could decide to strip the land of its timber, and small users were powerless to stop it.

In the course of the modern era, most northern European governments sponsored afforestation and tried to rationalize the use of wood. The governments of eighteenth-century Sweden, which included Finland until 1809, attempted to find ways to limit and rationalize the use of timber resources. Government officials found inefficient stoves, slash-and-burn cultivation, and tar production particularly wasteful. Already in the seventeenth century, the Swedish Riksdag tried to limit the number of sawmills in the country in order to spare woods that were needed for charcoal burning. Scottish iron masters in search of a safe supply of wood stimulated the planting of coppice fields. The idea that afforestation might in the future solve fuel shortages gained adherents in the Netherlands of the late eighteenth century (see the case study "Fuel Resources and Wastelands in the Netherlands around 1800" on pages 165–175). Similar developments took place elsewhere. The British Royal Society asked for a study on the causes and effects of deforestation as early as 1662. Many small states in Germany tried to be self-sufficient in wood for military reasons. Even in France, where wood was relatively abundant, the absolutist state imposed a sweeping forest ordinance in 1669. The afforestation movement itself matured only in the nineteenth century; in northern Europe as a whole, at the end of the early modern era, the state of many forests was precarious at best. Regions such as the Siegerland in western Germany, where sustainable agriculture, forestry, and industry were combined, were exceptional. However, there were strong signs that Europeans were learning from past patterns of resource use and that afforestation was squarely in the interest of states.

Wildlife

The history of European wildlife is a story of human selection operating through land management, prejudices, interests, and experience. Deforestation interfered much with animal populations, but what this process meant exactly for wildlife is not clear. Disturbing forests by cutting wood, harvesting acorns, or gathering leaves, grasses, or humus can be grasped in a simple model. The impact of such activities grew when forests diminished. When, as often happened, forests were parceled into isolated patches, species became more vulnerable to extinction than they would in a large ecosystem. According to Andrew Dobson, over the past fifteen centuries, approximately 0.1 to 0.3 percent of Europe's nature has been transformed into humanized habitats each year. This happened far more slowly than in the United States between 1600 and 1900 (about 0.7 percent per year) or today in the tropical rainforests (1 percent per year). Actual extinctions in Europe's agrarian history (before the onslaught of agroindustry) are rare, though many species were driven to the margins of society. To make matters more complicated, slow transitions in land uses—from virgin forest to different kinds of arable land or more intensively used woodlands—could allow greater biodiversity in the long run. In the Netherlands, for example, the landscape became far more diversified from the Middle Ages up to the nineteenth century, and greater biodiversity also resulted as more species of meadow-nesting birds like pewit and redshank thrived in man-made environments.

Moreover, dwindling biodiversity in northern Europe cannot easily be attributed to deforestation alone; ferocious hunting, poaching, and wanton extirpation of all kinds of animals took their toll. Squirrels, coveted for their furs, did not entirely disappear but became so scarce that in medieval times rabbits were introduced to take their place. Wolves were deliberately chased out of western Europe by governments, and beavers, boars, and the European bison could by the end of the early modern era be found only in the far north and east and in mountainous areas, at the fringes of civilization. Chasing wild animals was motivated not only by a search for safety or protection but also by the qualities attributed to different species in the Bible and in literature, qualities having little to do with biological or ecological reality. Stories about man-eating wolves were not completely fanciful, but fairytales exaggerated these fearsome images. Boars were considered lazy and stupid, squirrels thrifty, and foxes unreliable as well as smart. The latter animal enjoyed the benefit of the doubt, as he could change from a chicken-thief to a kind of Robin Hood, mocking kings as in the fables of Lafontaine.

Early modern Europeans could be fanatic hunters, and they developed many new breeds of dogs to help with the hunt: dachshund against badger, retriever against duck, foxhound against fox. The European elite favored such species as deer, wild boar, pheasant, partridge, heron, duck, and rabbit, pursuing them for their meat or fur, or perhaps as trophies to glorify the hunter. Other species, such as eagle, wolf, and lynx, were wiped out because they were considered too dangerous to live in proximity to humans. Foxes might harm poultry and were considered a prestigious catch. Otters and ospreys were thought to harm fishing grounds and were vigorously hunted. Smaller predators, such as badgers, weasels, martens, ermines, and many birds of prey succumbed to merciless hunting for feathers, trophies, eggs, and fur—or just because they were there. The great auk, a flightless seabird that lived in Scotland and the Orkney and Shetland Islands, became extinct because sailors robbed the eggs. Tiny animals such as frogs, snails, and moles also courted decimation at human hands. Though they were hardly dangerous, many Europeans cast them as outsiders in a human-dominated creation, therefore to be killed at will. Europeans could practice extreme cruelty toward "soulless" animals. Protestant churches opposed cruelty, but only in the course of the eighteenth century did caring for animals become more widespread in educated, bourgeois circles.

Governments opposed wanton cruelty as a pastime, but they, too, were prone to support the extermination of all kinds of "vermin," from the smallest mouse to the largest eagle. Following the devastating Thirty Years War, wild nature regained ground in depopulated areas of Germany, and officials promptly set bounties on wild animals. In 1697 the sovereign of Brandenburg published a list of animals with a price on their heads: boar, fox, otter, badger, wolf, marten, wildcat, weasel, owl, vulture, crow, eagle, raven, and even stork. Similar lists existed in sixteenth-century England and the Dutch republic, where many predators—eagle, buzzard, hawk, fox, and weasel—stood in the way of farmers and hunters as well.

Hunting for sport and prestige demanded specific manipulations of wildlife, the landscape, and the economy. To provide visibility and access to horses, woodcutters removed undergrowth and low branches in the forests. Hunting herons with falcons required wetlands, as did the use of duck decoys. Foxhunting in England called for a specific landscape of hedges and other obstacles to make it more enjoyable. Hunting was most easily tolerated where it did the least damage to economic activities. Large tracts of sparsely populated areas such as moors thus made excellent hunting grounds. But as wildlife did not recognize human borders, hunters inevitably demolished farmland and earned the enduring hatred of peasants. Their resentment echoes throughout the *cahiers de doléances* of 1789. Some countries, like the Netherlands, imposed rules to keep hunters at bay until after harvest. Many countrymen, however, wanted to hunt themselves; they complained less about hunting as such than about it being the privileged business of nobles and aristocrats.

Mining

Mining was essential to early modern economies, and regions or countries with large metal or coal deposits profited. The early modern era has rightly been called the "age of wood," and the age of coal and iron began only in the late eighteenth century. Nevertheless, metals were indispensable for ploughs, money, weapons, tools, and supports in wooden constructions such as buildings, ships, and mills. Over several centuries, coal took over the functions previously assigned to firewood and charcoal (see chapter 5 for a more detailed account).

The most important metallurgical industries grew around existing cores: from the south of Wales to the Midlands in England, in ever-wider circles around Liège in the southern Netherlands, the region east of Lyon in France, the southeastern German states, and the area to the northeast of Stockholm. Yet small metallurgical industries were widely scattered, depending on local ores, local transport facilities, and local woods for charcoal. Metal production rose steadily, with iron production tripling in Europe between 1500 and 1800. With the important exception of silver, exhaustion of mineral resources seemed remote, although imports from the Americas, especially silver but later iron and copper as well, were not negligible.

Apparent awareness of environmental problems occasioned by economic development shows in writings about both forestry and mining. Some treatises dwelled at length on the (environmental) pros and cons of mining. The most famous of these studies were those of Georg Agricola, *De re metallica*, published in 1557 and the most outstanding work on the subject until the end of the seventeenth century. In his first book Agricola, a doctor, analyzed how mining led to the destruction of woods, the digging up of fields, and the disappearance of birds and animals. Poisoned rivers caused fish to die, and working in the mines was unhealthy and dangerous due to the bad atmosphere and collapses. Morally damaging was the greed stirred by the digging of gold and silver. Agricola also took pains to argue in favor of mining, pointing out that mines were mostly dug into (purportedly useless) mountains, and that where forests had disappeared, arable land could be developed. Gold and silver could be used to buy fish, birds, and animals to repopulate areas formerly used for mining. For Agricola, the fruits of the new lands compensated the losses of woodland by a large margin.

The irrefutable argument was, of course, that the economy could not do without metals to make money and tools.

Fishing

The importance of fishing on the high seas was a new development of the sixteenth century. Medieval techniques for drying cod and curing herring made it possible to store fish, such that dried and salted fish could then become staples in certain regions, especially as high grain prices forced down meat consumption in the course of the sixteenth century. The Dutch famously exported herring to the Baltic states. Compared with other continents, Europe has a proportionally very long coastline, so fishing was a relatively important economic activity in all western and northern European countries. Despite the fact that the adjacent seas have the most abundant fishing grounds in the world, the threat of overfishing loomed early on a local scale. Along Holland's former inland sea, the Zuider Zee, different cities contested one another's rights to fishing grounds. In the fifteenth and sixteenth centuries, severe conflicts over cod-fishing near Iceland were fought out among England, the Hanseatic League, and Scandinavian kings. The discovery of seemingly inexhaustible stocks of cod in the northwest Atlantic, thanks to the development of more robust fishing boats, quieted these conflicts only temporarily. The Iberian countries, England, and France all had important stakes in the cod fisheries. From the mid-sixteenth to the mid-eighteenth centuries, about 200,000 metric tons of cod were caught in the north Atlantic. By the late eighteenth century this number may have been doubled, probably leading to local or regional overfishing.

Europeans, then, practiced unsustainable fishing well before the development of steamships. Some examples of sustainable fishing can be identified as well. Until the middle of the nineteenth century, the Dutch regulated the length of the fishing season, fishing techniques, and the size of catches along the shores of the North Sea.

Fishing was certainly not for free in Europe's more populated areas, for nations, provinces, lords, and cities claimed fishing rights on the seas, rivers, brooks, and lakes. During the early modern period, inland fishing grounds diminished as marshes and lakes were drained and rivers were canalized and, in the neighborhood of large cities, polluted. Some thirty species of fish in the Thames disappeared or became rare, among them bream, trout, and flounder; these species have since recovered in England. Polluted waters in central Europe, caused for instance by the retting of flax or mining, damaged fish stocks. It may be impossible to quantify the damage, but it is clear that some well-known species, such as salmon and sturgeon, were already becoming very rare. The practice of cultivating fish in ponds, though it declined in the seventeenth and eighteenth centuries, compensated to some extent.

Erosion

Overexploitation of the land was one sign of environmental stress in northern Europe in the eighteenth century. The law of diminishing returns was in full swing in many places as stories about erosion, failed land, clearances, and the appearance of sand drifts became commonplace. Many people had to cope with the problem of erosion. Soils erode quickly following deforestation, but the rate of erosion remains high for a long time thereafter, especially on steep slopes. Landslides and the silting up of rivers were the most striking consequences of eroded landscapes. In the second half of the eighteenth century, the French government promoted clearance of wastelands—a potentially harmful undertaking—in the name of economic development and physiocratic ideas: these emphasized the importance of land, not trade or industry, as the basic source of the wealth of nations. The local intendant of Roussillon in fact forbade further clearance because of soil erosion. Alsatians complained much about land erosion in their *cahiers de doléances* of 1789. Driven by population growth, soil erosion may have reached its peak in both France and Germany in the late eighteenth century. Ever more pastureland was put under the plough, causing the disappearance of sheep and cattle herds and with them the most important source of manure. The famous introduction of clover was not a panacea, at least not in Denmark and Norfolk. Compromised soil fertility and overexploitation were joint problems even in some of the most prosperous regions of northern Europe.

In extreme cases, overexploitation revealed itself in the enlargement and diffusion of sandy plains and dunes. Periods of severe drought or storms could accelerate their formation. In Denmark as everywhere, this phenomenon was caused by heavy deforestation and overgrazing, effects of both military buildup and population growth. As much as 5 percent of Jutland was covered with sand dunes in 1750. A contemporary account by a civil servant responsible for inspecting the sand dunes reads: "From several miles distant it appeared like a heath fire, way up in the air. The inhabitants came toward me, weeping, showing how their property had been destroyed." Comparable stories are known from the Netherlands, England, Sweden, central Europe, and the Atlantic coast of France. Some of these dunes had formed naturally, but even then, overgrazing and deforestation at the edges of the dunes could extend them. These sands represented the final stage in a slow process in which people impoverished the forest to shrubs, heath, and finally bare, sandy soils. Despite the fact that in some regions local governments concerned themselves with this process, only in the nineteenth century was it reversed due to large-scale afforestation (see the case study "Fuel Resources and Wastelands in the Netherlands around 1800" on pages 165–175).

Another form of erosion (easily overlooked by historians) developed on roads and their verges. Many Europeans were frequently on the move, locally, regionally, and internationally, and walking was for most people the only means of transportation. Not only male laborers but also young women and girls, looking for jobs as servants, moved from village to city. As a result, dust was one of the most common forms of air pollution. Because most road surfaces were sandy, holes developed easily. By circumventing these holes, carriages and carts widened roads considerably. More damaging could be cattle driving. The late medieval rise in meat consumption (see chapter 3) caused the drives of thousands of beasts from Scotland, Denmark, Hungary, and Poland, accelerated in the sixteenth century with the speedy growth of London and Amsterdam. Cattle herds formed tracks extending tens or even hundreds of meters in width. Clouds of dust on the horizon announced their coming. Dust also came with armies or the tens of thousands of seasonal, agricultural laborers who walked every year from Germany to the shores of the North Sea. Paved roads, canals, and railways would in due time solve this age-old problem.

AGRICULTURAL REVOLUTION

The early modern economy danced to the rhythm of population pressure. Grain prices particularly followed long-term fluctuations in population growth. In the sixteenth and early seventeenth centuries grain prices were steadily on the rise and were an important condition of the so-called price revolution. The combination of poor weather stemming from the Little Ice Age and strong population pressure could have devastating effects. Grain prices dropped between about 1650 and 1750 and rose again in the second half of the eighteenth century. Grain prices influenced such economic processes as land clearance, the choice between pasture and arable land, the demand for manufactured products, and the introduction of new crops or hybrids. Low grain prices stimulated higher demand for meat and wool, whereas high grain prices led to the opposite.

In the long run, feeding a growing population was one of the hardest won accomplishments of the early modern era. England and the Low Countries succeeded most in this respect. The number of persons that could be fed by 100 people working in agriculture was constantly rising. In England the figure rose from about 130 in the early sixteenth century to nearly 250 around 1800, and in France the estimates show an increase from nearly 140 to 170. In the Netherlands the already high number of the sixteenth century, about 175, rose to 250 in the same period. In analyzing the yield ratio—that is, the ratio between harvested seed and the amount of seed needed for sowing—B. H. Slicher van Bath distinguished four regions in Europe between 1500 and 1800. In the most advanced region, consisting of England, the Netherlands, and probably northern France, the yield ratio was about 7.4 at the end of the Middle Ages and rose, after a temporary regression to 6.7 in the seventeenth century, to a yield ratio of 10 at the end of the eighteenth century. In the second region, composed of southern France, Italy, and Spain, the yield ratio did not rise above 7. In the third region of Germany, Switzerland, and Scandinavia, the yield ratio barely rose above 5 in the second half of the eighteenth century. In the fourth region, eastern Europe, it did not even reach that figure. Given that the yield ratio dropped in periods of low population pressure, such as the seventeenth century, it is clear that it could fluctuate according to demand. The Agricultural Revolution was a movement strongly stimulated by population growth. The yield ratio also makes clear that major improvements in agriculture were mostly restricted to the first region and some parts of northern France. The Agricultural Revolution was based on the introduction of already known, late medieval routines from Flanders and Holland to England, supplemented by new ideas. It did not stem from a few revolutionary ideas, such as planting clover to add nitrogen to the soil, using the seed drill invented by Jethro Tull, or enclosing common lands. Enclosures were more common in seventeenth-century England, before the onset of the Agricultural Revolution, than in the eighteenth or nineteenth centuries.

The enclosure movement in England did, however, alter the landscape, which became less open. As commons were often wastelands, the enclosure movement further circumscribed "nature." The same had happened in France, where more than 100,000 hectares of wasteland were cleared in the late eighteenth century. A Finnish case study shows that in that country common lands, used for slash-and-burn farming in the early nineteenth century, became private property, driving the landless off the land, deeper into the forests and into employment in the ironworks.

For the ordinary farmer the Agricultural Revolution was a slow and piecemeal fight against the odds, learning by doing in seeding, breeding, manuring, and introducing new crops. The remarkably successful introduction of the potato made it possible to feed three times as many people as with wheat from the same acreage. Central to the development, however, was the relative freedom of farmers to try out something new. Better transport facilities, such as better roads and canals, served farmers well, and grain imports from eastern Europe acted as a buffer against disasters. The Malthusian gap between population and resources was further closed by the draining of marshes and land reclamation. The Dutch became most famous for this all over Europe. In England, France, central Europe, and Sweden, *Hollanderies* (a French word for "polder," meaning a low-lying area reclaimed from a marsh, lake, or sea, and enclosed by embankments) added considerably to the stock of fertile land. Not all these projects were successful, however, and in Holland itself, most polders were created in the seventeenth century; with falling grain prices after 1650, the process lost momentum.

Notwithstanding the accomplishments in agriculture, one should be skeptical of optimism concerning rural living standards. Major peasant revolts were absent during most of the eighteenth century in northern Europe, but in many areas, and not only in France on the eve of the Revolution, pauperism was widespread. About one-quarter of the relatively well off Swedish peasantry were cottagers and paupers around 1750, and about one-half were cottagers and paupers in 1815.

Small wonder that when the frontier in the United States started to move west in the nineteenth century, many Europeans decided to emigrate. In the end, the most important additions to the European search for means of supporting the growing population came not only from the Agricultural Revolution but also from overseas, from the European "ghost acreage."

GHOST ACREAGE, OR ENLARGING THE ECOLOGICAL FOOTPRINT

Malthus was troubled by the limits the land placed upon population growth. However, from a global perspective, Europe's economic potential was not dictated by diminishing returns, but rather by the opposite. The early modern world was an expanding one. The discovery of the Americas notably widened the scope of economic and demographic possibilities. Diseases, war, and maltreatment decimated Native Americans, reducing their populations in the whole hemisphere by 90 percent within two centuries of initial contact, from a high of at least 75 million but possibly as many as 145 million. The Americas nearly became "empty" space. New uses of the high seas and the opening up of Russia were other dimensions of European expansion. For Europeans the discoveries resulted in a most welcome boosting of resources. Without them, a Malthusian trap might have been a real threat.

A concept developed in historiography is the notion of "ghost acreage." Georg Borgstrom introduced it in 1972 in his study *The Hungry Planet*; Eric Lionel Jones later applied it to history in *The European Miracle* (1981). The concept has affinities with the idea of an "ecological footprint" (see the Glossary). In the case of early modern Europe, ghost acreage was the additional increment of "land" added to Europe because of fishing and whaling in the oceans and because of forestry and agriculture in the Americas. For example, whales were hunted for oil that was used for lamps. Because of whaling, land in Europe became available that otherwise might have been planted with hemp or rapeseed. Basques overhunted whales in the Gulf of Biscay already in the early fifteenth century and extended their hunting grounds to the Atlantic Ocean. Later on, the Dutch, English, and Norse joined the hunt. In the seventeenth and eighteenth centuries, whalers were active around Spitsbergen and Greenland until they had severely depleted north Atlantic stocks; whalers were forced to move south in the following century. Between 1669 and 1800, more than 100,000 whales were killed in the north Atlantic, diminishing the population of Greenland and Right whales (which grew about 2 percent per year) from roughly 80,000 to 50,000. Despite strong signs of overexploitation later in the eighteenth century, those catches were very moderate compared with those of the nineteenth and twentieth centuries. Besides, they involved just two species of whales, both of which had the advantage that they did not sink after being killed.

The importation of cotton is another example of the idea of ghost acreage. The British cotton industry expanded rapidly in the second half of the eighteenth century, with production rising tenfold between 1760 and 1785. Mill owners imported cotton on a very large scale, first from India, and in the nineteenth century from the United States, thereby saving land that otherwise would have had to be used for sheep raising or growing hemp or flax. Britain would have needed another 9 million acres of sheep pastures in 1815 had it grown its own woolen fibers instead of using cotton. Timber imports from Scandinavia and later the United States saved much land also, as did sugar imports from the West Indies. If furs had not been imported from Siberia and North America, Europe would have had to set aside acreage for hunting, or for raising animals like rabbits, in order to consume the same number of furs. North America exported on average 150,000 predator and beaver furs to England and another 260,000 to France per year between 1700 and the end of the Seven Years War (1763). Russia exported beeswax, honey, seal oil, hides, and tallow to western Europe. The enormous quantities of cod, imported from northwest Atlantic waters, added food without which large tracts of land would have been required to raise livestock to obtain the same amount of protein. The high seas, the Americas, and the sparsely populated areas in Europe's north and east ap-

Artifacts from a sixteenth-century Basque whaling station, Canadian Museum of Civilization. (Corel Corporation)

pear as an "unending frontier." Europeans could not possibly, in fact, have set aside acreage at home to match the imports in equivalencies. The "world hunt," as John Richards calls it, with northwest Europe as its base, was inaugurated in early modern history and led in the long run to very serious depletions and overexploitation of nature all over the globe.

Coal also saved land because less woodland was needed. Around 1780 there was still enough coppice for the English iron industry, so the early phase of industrial takeoff was possible without coal. It has been calculated that an impossible amount of woodland would have been needed if iron producers had continued to use charcoal by the year 1850 (see chapter 5 for a more detailed discussion). This counterfactual argument does not imply, however, that an industrial revolution would have been impossible without coal, but rather that Britain would not have been the most appropriate place for it to begin. Sweden had an iron industry as well and harbored tens of millions of hectares of forest, not to mention the forests of neighboring Finland and Norway.

To complicate the picture, English people were not always happy with these developments. They considered cotton clothes inferior to woolens, preferred stone over brick, and detested polluting coal smoke; the rich long delayed

Native Americans aboard ship to exchange furs for European goods, seventeenth century. (North Wind/North Wind Picture Archives)

changing to the new but inferior energy source of coal for cooking and heating. Moreover, producing the new substitutes required more work, and longer working hours led to what has been called the "industrious" revolution. In the words of Richard Wilkinson, the industrial revolution "should not be regarded as the outcome of a society's search for progress, but as the outcome of a valiant struggle of a society with its back to the ecological wall," a struggle English society was well aware of. (Wilkinson 1988, 91) The point may be exaggerated, for the English envied the Dutch in the seventeenth century for their smoothly operating, canal-based, comfortable public transport system, and the use of bricks surely meant better and safer housing. One could argue that society did not have its *back* to the ecological wall but was facing it, for the old ecological order could hardly provide higher living standards for the masses. The Malthusian trap did not close, but without ghost acreage, another subsistence crisis would have been a real possibility in the later eighteenth century.

URBAN POLLUTION

In his famous novel *The Perfume*, Patrick Süsskind gave a clear idea of the stench in eighteenth-century cities. In Amsterdam, the expression "cucumber time" described the annual moment when the financial markets came to a standstill. At the time when cucumbers happened to be harvested in the summer, the stench of the canals became so unbearable that the regents and rich merchants fled the cities to their summer residences.

Most urban stench was caused by all kinds of rotting: bones, rotting vegetables and fruits, leftovers from food and fish. Feces and urine, from animals and people, also contributed to the situation. The historical literature on pollution in preindustrial cities abounds with horror stories, such as men defecating on rooftops or whole carcasses thrown into canals or rivers. The death rates in the cities were abnormally high, higher than the birthrates; cities grew only because of immigration from the countryside, where the situation was somewhat better but employment opportunities were scarce.

Cities in northern France were even less healthy at the end of the eighteenth century than during the Middle Ages. Medieval cities relied on running water from wells and rivers. Watermills produced energy for industries such as grinding and fulling. The woolen industry, too, depended on clean water, and dyestuffs were not very polluting. The same could be said about curing and tanning in the leather crafts. Bathing, though restricted to the elite, was seen as healthy. Towns built sewers, and pigs and geese roamed and cleaned the streets.

In the course of the early modern era, urban water became stagnant. Due to frequent warfare, starting with the sixteenth-century wars of religion, cities began to protect themselves with moats and ramparts. Windmills replaced watermills. With the coming of linen, which to a large extent replaced wool, new dying techniques and processes such as bleaching and retting were introduced; they required stagnant waters and a humid environment. In the curing of hides, animal excrement replaced alum. Stagnant waters meant more infections, more stench, and more malaria, a common disease in modern times all over Europe and extending about as far as Archangel, a city near the mouth of the Dvina River in northwest Russia.

In England, "modern" air pollution arrived because cities, especially London, switched from fuel wood to coal. Sea coal from Newcastle, with its high sulfur content, could be blamed for much pollution. Smoke and sulfur dioxide concentrations rose constantly in London from the end of the sixteenth century until the end of the nineteenth century. In 1578 Queen Elizabeth complained about the shocking smells in London. Margaret Cavendish narrated in 1660 how her husband was moved when, after years of exile, he discerned the smoke of London on the horizon. Statues in the city were already blackened in 1750. Smoke, stench, and noise made early modern cities unpopular among writers and intellectuals, who began to glorify the blessings of country life. The "country" itself entered the cities as pigs, poultry, and even cattle were held inside the cities well into the nineteenth century. London in 1866 still harbored more than 9,500 milking cows.

Case studies from the territory that would become Belgium in 1831 show us that many other industries and crafts contributed to the growing load of urban pollution. The production of meat, beer, ceramics, paint, bricks, paper, and gin all created nuisances affecting the air, water, and streets in their environs. As cities had a long tradition of government interference, we can trace many judicial cases involving requests, regulations, and complaints about environmental issues. Citizens complained about certain trades, and various branches of industry had problems with one another, as well. Though it is not easy to know what the effects of the regulations were, a wide range of measures existed to preserve the peace. Cesspool workers were obliged to work at night, but not in the summer, and they had to transport excrement (which formed excellent manure) in closed barrels. In Antwerp bleachers were protected from fumes emanating from other industries, which were not allowed to establish themselves in the same street. Oil crushers were not allowed to work at night because of the noise, whereas candle makers had to work at night because authorities hoped that complaints about smells would lessen. One important principle operated: cities had to balance employment and health, and they did this by separating disturbing or polluting activities in time and space. A fine example of this was the fate of smithies in Bergen in Hainault, who were put on the spot in the eighteenth century by city authorities: they had to either move out of town or limit their working hours. Urban authorities tended to interfere little with technology, and outright prohibitions of crafts were unusual. More active policies were germinating, however. Assuring the quality of fresh water, building sewers, and, in general, fighting stench would become urban obsessions in the nineteenth century (see chapter 5 for a fuller discussion).

DISASTER MANAGEMENT

The northern European population suffered intermittently from the disasters of hunger, war, fires, epidemics, storms, floods, and avalanches, to name only the most frequent events. Some of these disasters were rather famous in their time. An avalanche in Switzerland in 1618 that caused 1,500 deaths has been recorded as one of the worst weather-related disasters. The Great Fire of London

Londoners flee the Great Fire of 1666 in boats along the Thames River. (Getty Images)

in 1666 is another example of a famous catastrophe, but Scandinavian towns were the worst hit by fires overall. Storms damaged harvests and wrecked ships. Plagues killed men and animals. In the seventeenth century, smallpox raged in the Netherlands and the plague in London. Rinderpest killed millions of cattle in the first half of the eighteenth century.

States and urban governments experimented with regulations and measures to alleviate the effects of disasters. Quarantine, based on the biblical notion that forty days were necessary to outlive a crisis, proved to be a reasonably successful way to limit epidemics, provided it was strictly enforced. Major outbreaks became scarce in the eighteenth century. Massive slaughter and bans on the movement of herds helped combat epizootics. Eric Lionel Jones speaks in this respect of the bovine "cordon sanitaire" of French troops on the border to prevent the passage of contaminated herds during the 1740s. Governments, for example those in London and Prussia, sometimes ordained the use of bricks instead of wood to lessen the risk of fire. The French government of Louis XV successfully redistributed grain to counteract famines. Lighthouses to prevent the wreckage of ships also became more common. Such developments were not without setbacks. Government officials worked inefficiently and could be corrupt. Governments were, moreover, not only concerned for the well-being of their people. Disasters undermined the economy, could develop into social protests (especially in the case of famines), and diminished income from taxes. Those taxes were needed to pay troops, who became better nourished and maintained. Following the havoc wrought by the wars of religion in the sixteenth century, the Thirty Years War in central Europe, and the wars of Louis XIV, eighteenth-century wars were less disastrous. Feeling responsible for the well-being of the people slowly became customary for state and city authorities in Europe.

THE KNOWLEDGE REVOLUTION

For sixteenth-century Europeans, the world was full of magic. Before the Protestant Reformation, saints and their relics, alchemy, the zodiac, and evil spirits influenced health and wealth. It was a world full of omens such as the appearance of comets. Though not strictly Christian, these beliefs were widespread. Especially in the Protestant countries of northern Europe, the Reformation led to a strong offensive against occultism and saint cults, which played a tremendous part in what Max Weber called the *Entzauberung*, or disenchantment, of the world. Superstition did not wholly disappear with the Reformation, however. A backlash in the form of fanatic witch-hunts did not spare the Protestant countries and died out only in the second half of the seventeenth century.

Popular views of nature oscillated between superstitious fears and trust in a God-given chain of being. Dominant was the idea that Man, though lower than the angels, was at the center of the universe. He was condemned by the fall to till the earth, but he had divine permission to use nature as he saw fit. Copernicus and Galileo struck an initial blow at this idea, for their astronomy shifted the center of the universe away from the earth. Moreover, rational thinking introduced a mechanistic worldview, though one not opposed to the Bible's account of creation for the time being. Little by little, worldviews became far more sober and nature more comprehensible. This more rational approach to reality was one of the pillars supporting the remarkable spread of knowledge in the early modern era. As a reaction to this mechanistic worldview, the eighteenth century witnessed the early Romantic movement with its awestruck sense of nature.

The relationship between science and practical or tacit knowledge is hard to pin down, so the term "scientific revolution" has been widely discussed and criticized. For that reason, historian Joel Mokyr prefers the term "knowledge revolution." It was a process strongly stimulated by the invention of the printing press in the late fifteenth century. With time it became more customary to see all kinds of practical knowledge appearing in books. The most famous example is the eighteenth-century French *Encyclopédie*. The art of visual representation and mechanical illustration was older and already well established in the sixteenth century, as exemplified by the work *De re metallica* on mining. Geography and the art of drawing accurate maps were strongly rooted in Europe. Other intellectual domains that began to shape the environmental history of this period include meteorology, agricultural science, gardening, forestry, and medicine.

In the early modern era it became more common to describe and measure weather conditions, and for that reason it is possible to reconstruct the history of climate during the Little Ice Age in part by relying on primary sources. A most remarkable example is the work of the Swiss Wolfgang Haller, who made more than 10,000 weather observations between 1545 and 1576.

In agriculture, gardening, and forestry, voluminous information was collected, systematized, and shared by way of lectures, scientific organizations, travel, and the publication of books. The travels of Arthur Young, the great publicist of the Agricultural Revolution, are still famous as an example of hands-on investigation of agricultural practices. Cataloging botanical information and creating botanical gardens, begun during the Middle Ages in cloisters and among kings and nobles, developed into a more sophisticated science of its own in subsequent centuries. Germans, from the sixteenth century onward, pioneered scientific forestry. Early "foresters" began by translating antique writings, but they soon added their own experience. In Germany silviculture developed in the eighteenth century against the background of a growing demand for wood and the mercantilist ideas of German princes. In France, the ideas of the physiocrats had a profound but not always positive effect on forestry. They promoted laissez-faire and successfully agitated against Colbert's Forest Ordinance of 1669. The force of their ideas, in combination with rising population pressure, led to increased logging in high forests.

Impressive progress was made in medicine, against the odds, but the effects only slowly became visible. An early example is the sixteenth-century medical writings of physician and alchemist Paracelsus, whose dictum that "the dose makes the poison" is still in use. The seventeenth-century Dutch biologist Antoni van Leeuwenhoek made the remarkable and interesting estimate that the earth could support a maximum of 13.4 billion people. Others worth mentioning are James Lind and his study of scurvy in 1746, and the Italian Giovanni Lancisi, who in 1711 described the relation between malaria and mosquitoes. Finally, we should mention the work of Edward Jenner on vaccination, which appeared at the threshold of the nineteenth century. These modern ideas should not hide the fact that, until well into the nineteenth century, actual health services were still mostly based on awkward theories around quarantine, bloodletting, and miasmas. Nevertheless, in the early modern era, reason-based knowledge became, for better or for worse, a legitimate basis of Europeans' interactions with nature.

DENYING MALTHUS

The demographic situation in early modern Europe, before modern hygiene and improved living standards prevailed, has been compared with a room where many people are knocking on the door to come in (high birth rates), and many people are going out at the other end (high death rates), leaving the room itself rather empty. Once death rates began to decrease, many people continued to come in, but fewer people were exiting, making the room more crowded. This happened first in the sixteenth century, creating the famous price revolution, lowering living standards, and stimulating agricultural expansion. Later in the seventeenth century, however, a Malthusian crisis was far off due to low population pressure. As a result, yield ratios diminished everywhere in Europe, but by the end of the eighteenth century the threat of dearth, even famine, structured a majority of human lives. Population growth drove many Europeans to the margins of subsistence. By that time, the people in northern Europe lived, from an environmental perspective, in an impoverished world. In comparison to the late Middle Ages, Europe was home to less wildlife and fewer forests. Climatic conditions had become unfavorable. The landscape showed severe signs of erosion. In the cities, lack of knowledge about hygiene and the causes of epidemics and diseases, in combination with food scarcity and precarious employment, resulted in Rousseau's urban graveyard. Living standards had barely risen since late medieval times

But the Malthusian trap did not close as Europe acquired enormous "ghost acreage"—the high seas, Russia, and the Americas. In these regions a brutal "world hunt" set in from the later Middle Ages onward. Draining marshes, creating *Hollanderies,* and clearing wasteland added land as well. The Agricultural Revolution accomplished an important rise in productivity in the long run. For these reasons, the necessity of creating a sustainable economy was absent, even more so as coal mines started to relieve the expanding economy of severe energy shortages. Additional sources of energy gave economic actors more flexibility and more choice in the ways they could use land, and more energy turned afforestation into a realistic option.

At the end of the eighteenth century, despite severe social problems, life expectancy was somewhat longer, disasters were fewer, and wars less destructive. However, the most important long-term development in the environmental history of the early modern era was the combination of a surge in knowledge—about geography, resource use, forestry, agriculture, and medicine—and governing forces slowly taking more responsibility for economic development, health, and public safety.

5

NEW REGIMES OF PRODUCTION AND POLLUTION IN THE INDUSTRIAL AGE

In 1816, as countless communities in central, western, and northern Europe were picking up the pieces from a quarter-century of warfare, crisis of a different nature struck. The previous year had seen the end of the French Revolutionary and Napoleonic eras, with a structure for European peace hammered out at the Congress of Vienna. Momentous changes in law, governance, ideology, and warfare had cascaded well beyond the borders of France since the fall of the Bastille in 1789, yet a traditional, sometimes fragile agrarian base continued to shape life for the masses. Most livelihoods depended on the grain harvest, and the grain harvest depended on the weather. Harvests determined consumer purchasing power and the prices of raw materials for manufacturing. Events of the next several years tested the limits of the agrarian system; although no one knew it, Europe was about to face the final, widespread subsistence crisis in its history to date.

Halfway around our planet, a volcano called Tambora, located on the island of Sumbawa in present-day Indonesia, erupted in 1815. Twelve thousand people on Sumbawa died. Written testimonies prove that the explosions were heard 2,000 kilometers to the west. Volcanologists deem this natural catastrophe to be the greatest eruption since 1500 if judged by the volume of Tambora's ash, which reached a minimum of 150 cubic kilometers. Dust drifted in veils through the stratosphere, rendering the atmosphere less transparent, thus less able to absorb solar radiation. Surface temperatures consequently fell, and by 1816 the western hemisphere was experiencing some of the lowest spring and summer temperatures ever recorded. Harvests failed from central Europe to eastern North America; dearth and famine struck Europe in widespread regions from the Ottoman Empire to Ireland, most severely in Switzerland and the southwestern German states. Typhus and bubonic plague rode on the heels of undernourishment, and shortages lingered in some areas until 1819.

The subsistence crisis of these years provides a dramatic example of Europe's vulnerability to distant earthly phenomena, but a historical context also helps to explain the magnitude and geographical reach of the calamity. A postwar recession was already driving up unemployment just as several million discharged soldiers entered the labor market. Many families and communities were, in other words, defenseless against skyrocketing bread prices. Even so, Europeans were able to mitigate the food crisis, severe though it was, as they reaped a harvest of changes in transportation, commerce, and social institutions. First, Russian and North American merchants responded by exporting grain to central and western Europe, though not all European governments jumped at the opportunity to buy. Urbanized regions of the Low Countries had depended on grain imports for centuries, but improved exchange now freed nearly the whole continent from dependence on local food production. Second, thanks to eighteenth-century state-building efforts and rising humanitarianism, public and private charities acted as buffers more effectively than in the past. The result was that mortality did not peak as high as it had during the two worst subsistence crises of the eighteenth century (1709–1710 and 1740–1743), despite the resounding failure of the harvest. Europeans had become marginally more sheltered from the vagaries of nature through deepening connections, both geographical and social. The nineteenth century would see a partial transformation of the agrarian regime, a revolution in manufacturing and energy, and, perhaps most significant of all, an ideology that celebrated technical progress and liberation from natural constraints.

MODERN AGRICULTURE: INTENSIFICATION AND SPECIALIZATION

Agricultural productivity was already rising, due in part to technical change but even more to increases in cultivated acreage, the abolition of fallow, and specialization. Farmers consolidated the gains of the Agricultural Revolution (see chapter 4) in much of northern Europe. In Great Britain a final great wave of enclosures—some 2,800 Acts of Parliament between 1760 and 1850—reconfigured about one-third of the land of England and Wales into intensively farmed, rectangular fields, bordered by hawthorn hedges interspersed with trees. Farmers reclaimed nearly one million hectares of so-called "wastes"—moors and heath—for agriculture, but a much larger total acreage, around five million hectares of English lowlands, was drained in the course of the nineteenth century. A similar trend remodeled the Scottish lowlands and middle-altitude moors. The ecological consequences were mixed: whereas bird species found havens in the hedgerows (whose extent in Britain was doubled), waterfowl, fish, and insects suffered from drainage, such as the reclamation of the Fenlands of East Cambridgeshire, enclosed and drained via steam technology late in the century.

"High farming," as it later became known in Britain, combined heavy capital investment, new and imported fertilizers, and limited mechanization with more traditional crop rotations, animal manure, and the use of human and animal muscle for energy. In the age of industry, farming was "not preindustrial and not fully industrialized." (Winter 1999, 52) Steam power remained too expensive for most farmers, and steam-driven implements proved too cumbersome on Britain's small, enclosed fields. It was on the whole a highly sustainable agricultural system, although inputs from distant sources—such as oilseed cakes fed to cows—improved soil fertility in Britain while reducing it elsewhere. High farming provided an image of the countryside that would become an object of care and a standard of beauty for many British preservationists.

In the agricultural powerhouse of France, land reclamation had been state policy from the 1760s. A Napoleonic law of 1807 further encouraged the drainage of marshes and allowed the state to undertake drainage if private landowners balked. The heyday of drainage and digging irrigation canals covered the decades up to 1860, by which time nearly all shallow marshes in France had disappeared in the name of improvement. Sown with grass, the marshes became highly coveted meadowland. Fallow land, too, fell gradually to the plow as peasants adopted clover and lucerne to maintain soil fertility, and as root crops such as turnips, potatoes, and beets became more popular both for human food (providing more calories than grain per unit of land) and livestock feed. Better fed livestock then produced more manure to return to the soil. These were the key changes of the century that allowed French farmers to feed growing urban populations. Yet to a larger extent than in Britain, small, owner-occupied properties, local resources, and common access structured agriculture in France. A host of traditional collective rights—to glean in harvested and unenclosed fields, to pasture in common meadows, and to cut wood in forests—had been guaranteed by the Rural Code of 1791 and remained fiercely defended in the nineteenth century. Still, the great property shuffle during the French Revolution had in many regions benefited the urban bourgeoisie, for whom productive land remained the best investment for decades to come.

In the German states, legal change helped drive the evolution of agroecosystems. During the first half of the nineteenth century, most German states reformed the structure of the so-called agrarian constitution by formally liberating peasants from their dependency on feudal lords. Lords profited most from these reforms, however, in that peasants owed them financial compensation. As peasants took out loans, the former lords accumulated liquid capital and bought up land. They created large agricultural enterprises, especially in the northeast of Germany, combining fields that they had once parceled out. Geared toward market requirements, agricultural production soared: whereas in 1871 Germany produced 23 million tons of potatoes, the crop rose to 50 million tons in 1912. Financial investments enabled mechanization and the plowing up of uncultivated land.

Before midcentury, larger landowners throughout northern Europe were extending monocultures and abandoning fallow periods. Distinct grain regions, vine regions, and cattle regions emerged. In France, intensive pastoralism claimed a broad swath from the Atlantic departments through Normandy and much of the north and east, whereas Languedoc in the south was remade into a vast monoculture of wine grapes. Monocultures were beginning to show their fragility, as demonstrated by the ravages of phylloxera in late-nineteenth-century France, but continuous rotation protected soils from erosion, and regional specializations did make better use of the natural potential of given areas. Farmers explored that potential in news ways, because land was becoming a productive unit whose profitability had to be maximized. The first large German monocultures came into being in the regions around Magdeburg and the Lower Rhineland during the 1860s. These areas became important suppliers of sugar in Europe, with sugar beet replacing imported sugar cane. Farmers cultivated sugar beet on gigantic fields, and entrepreneurs opened sugar refineries in their vicinity. The German sugar industry provides an early example of commercial monocultures' dependence on technical improvements, capital, and proximity to infrastructure and markets.

The high productivity of northern European agroecosystems grew to depend on fertilizers—nitrates, phosphates, and potassium—first extracted from minerals and industrial residues, and ultimately synthesized at the end of the nineteenth century. Swedish chemist Johann Gottschalk Wallerius invented the concept of agrochemistry in 1761, but German chemist Justus von Liebig provided its theoretical foundations in 1840. For Liebig, farmers required expert assistance in adapting to natural laws. The first publicly funded agricultural chemist in Sweden was a German, Alexander Müller, whose faith in chemistry led him to promote agriculture even in remote areas of northern Sweden. In 1918 another German, Fritz Haber, won the Nobel Prize in chemistry for perfecting his "Haber-Bosch synthesis" of nitrates using ammonia. By then northern Europeans could produce soil's three macronutrients through industrial processes. The scientific approach to nature carried strong nationalistic overtones; proponents in Sweden hoped to mobilize agricultural resources in the quest to reconquer Finland (under Russian domination since 1808), and Fritz Haber himself wished to reduce Germany's dependence on guano imported from Chile. His synthetic nitrates eased Germany's critical food situation for much of World War I.

All of these developments arose because urban markets for agricultural produce were expanding, both in absolute size and in terms of the radius from the producing areas. Faster, more efficient transportation systems expanded that radius; in Britain, for example, canal construction peaked in the 1790s, partly erasing the longstanding need for better roads. The next generation saw a national network of railroads constructed in two spurts during the mid-1830s and mid-1840s. French engineers carried out plans for a national rail system after 1842. It is worth noting just two of the many other ways in which country and city became more tightly connected in the nineteenth century: manufacturing enterprises invented new standards of profitability that farmers would have to reckon with, and urban populations became dependent on rural food production as gardens and food animals gradually disappeared from back lots and city streets. Widening markets also heightened competition among farming regions. It is no accident that the Industrial Revolution began in two agriculturally marginal areas of England—east Lancashire and the West Riding of Yorkshire. References already made to steam power and railways now demand a fuller exploration of the Industrial Revolution, ecologically the most significant European trend of the nineteenth century. As a global legacy, Europe's Industrial Revolution would be matched only by imperialism (the two went hand in hand) in overturning ways of life everywhere over the following two centuries.

ENGLAND'S INDUSTRIAL BREAKTHROUGH: MARGINALITY AND THE DEFORESTATION HYPOTHESIS

The hills of east Lancashire were good for pastoralism but little else: as would happen elsewhere in Europe, upland agriculture could not compete with that of the lowlands of southern and eastern England after the Agricultural Revolution. East Lancashire further resembled the uplands of continental Europe in that not all family members could be employed to raise livestock and workers had therefore turned to domestic industry. Textiles had come from rural Lancashire since the Middle Ages. This agriculturally marginal district had long been linked to markets and merchants through a system called protoindustry. What east Lancashire lacked in fertile soil it made up for in water, which was cheap and highly available along its upland streams. It was there that the first mechanized industry of cotton textiles—rural, decentralized, and dependent on traditional water power—developed following eighteenth-century technical innovations that increased the output of weavers and spinners. Likewise, the poor, pastoral, water-rich West Riding of Yorkshire was to become the home of a mechanized woolen industry.

Were these "pull factors"—the existence of a manufacturing infrastructure with its skilled workers and market connections, the improving transportation

A plume of smoke and steam rises from a train crossing a bridge in London, ca. 1885. Railroads drew country and city closer together and accelerated urban growth. (The Illustrated London News Picture Library)

systems, the growing markets—enough to make the Industrial Revolution happen? Ecological constraints, or "push factors," may also have played a role in England's industrial takeoff, although this remains a subject of controversy. On the one hand, a diminishing supply of wood and a shortage of agricultural land, relative to a rapidly growing population, provided a stimulus that helps explain the precise timing of the Industrial Revolution. Wood, the very basis of European material culture, was becoming so scarce in the largely deforested British Isles that the price of coal—known and used for centuries but usually more costly to extract than wood—grew competitive. Prior to the Industrial Revolution, smiths, lime burners, salt and soap boilers, dyers, and brewers used coal; even iron smelters began to convert to coke, a distilled form of coal, in the early eighteenth century, and the city of London was relying on over one million tons of coal per year for domestic heating prior to 1800.

Driven by wood shortages, heightened demand for coal called for deeper coal mines, which in turn stimulated the invention of increasingly powerful pumps to remove water from the mines. Thomas Savery's coal-burning, steam-powered pump, patented in 1698, served the purpose and was the first primitive steam engine, although it was quickly eclipsed by Thomas Newcomen's more efficient model early in the eighteenth century. Only when the mechanized cotton mills multiplied the demand for power many decades later did James Watt's innovations create a steam engine (first used in Nottinghamshire in 1785) whose linkages produced rotary motion efficiently enough to be used in manufacturing. Just as manufacturers favored coal over wood, transporters of bulk commodities favored canals over roads, another example of ecological constraint pushing change: roads required far more horses, and horses required substantial acreage to feed them, acreage in high demand for cereals as Britain struggled to remain self-sufficient in food production. Steam-powered locomotives, too, can be viewed as a substitute for horses.

On the other hand, economic historians have challenged the deforestation hypothesis of industrialization in particular by pointing out that wood prices varied considerably across regions within England, and that high prices were more characteristic of wood for construction rather than of wood for heating or charcoal. One historian estimates that British domestic industry might have survived profitably on wood and charcoal derived from less than 300,000 hectares of well-managed woodland. In this view, British manufacturers were faced with a choice of fuels, and the relative economics of extraction and transportation—not an absolute shortage of wood—gradually dictated the transition to coal. Such reasoning appears to suggest that demand for energy remained static. A leap ahead to the year 1850 reveals that the British were by then producing 2 million tons of iron annually; if they had used traditional charcoal-

James Watt studying improvements for the Newcomen steam engine in his laboratory, undated illustration. (Bettmann/Corbis)

powered forges, 20 million hectares of forest would have been required to produce the charcoal, an acreage one-third again the size of England and Wales. Even a late-eighteenth-century perspective, which is more historically relevant, shows the rising demand for energy across manufacturing sectors. Neither charcoal nor water mills could provide enough. Finally, although Britain imported enormous quantities of timber, mostly for shipbuilding, the price of imported wood was far beyond what manufacturers could afford to pay for fuel.[1]

Steam pumps allowed miners to dig for coal and metals at greater depths, and Watt's coal-burning steam engine resolved the energy impasse of the late eighteenth century. The economies of western Europe made a gradual but momentous shift from solar energy, captured mostly through an annual harvest of photosynthesis, to fossil-fuel energy—a form of solar energy captured over an extremely long period of time. Locally abundant coal supplies fostered rapid economic growth and produced a surplus of energy, creating a positive feedback

Manchester, England, with its textile and other industries, was the world's first great industrial metropolis. (North Wind/North Wind Picture Archives)

effect. By the early nineteenth century, the distribution of coal supplies partly determined the location of mechanized industry; the Borinage region of Belgium, the northern departments of France, and the Siegerland and the Wupper valley of what was to become Germany saw the first steam engines on the European continent from the 1810s onward. In England the textile industry continued to develop heavily in Lancashire, now thanks not to mountain streams but rather to the accessible coal seams of the Pennines. Steam engines and the power loom required economies of scale and larger labor forces than did water wheels and handlooms, so industry moved into towns such as Manchester, which was becoming a manufacturing metropolis by 1800.

Industry developed in parts of Scandinavia from the late eighteenth century. Denmark experienced an intensive phase of industrialization following upheavals in agriculture and commerce. The country's first steam engine began operating in 1790, and soon thereafter a textile industry acquired a base in Copenhagen. Competition from England and the depressive effect of the Napoleonic wars sent early Danish industry into decline. It remained in its infancy until around 1840.

SULLIED LAND AND AIR: THE IMMEDIATE ENVIRONMENTAL EFFECTS OF INDUSTRY

The effects of mining and coal combustion remain two of the many unfinished stories from the Industrial Revolution. To evoke just the English example, the "Lancashire System" of mining created a landscape pockmarked with pitheads, as coal entrepreneurs sought to limit the length of underground tunnels. Landowners, who under English common law possessed the mineral rights to their land, often negotiated leases that required damaged land to be restored for agricultural purposes; however, it frequently happened that landowners later sold their estates to speculators and mining entrepreneurs. Restoration of abandoned mines became mandatory only in 1943, in the era of open-cast mining. From the beginning, streams received unusable coal and the black water from washed coals, while floods brought coal dust that remained on fields and meadows.

Far from the pitheads, coal combustion accounted for much of the increase in particulates and chemical compounds that human activities spewed into the atmosphere. Paradoxically, large-scale industry did not cause most pollution in the early Industrial Revolution. Thousands of open hearths in individual households, often burning sulfurous bituminous coal with insufficient oxygen, sullied the air in Europe's teeming urban areas with particulates and sulfur dioxide. Domestic hearths produced 85 percent of all smoke in Great Britain; hundreds of thousands of fireplaces in London contributed to a gradual rise in the frequency of fog from about 1750 through the 1890s; a more apt term, "smog" (a word created by combining "smoke" with "fog") was coined in 1905. Noxious elements other than smoke contributed to the fogs: "Dust loaded with fecal matter, hot air, sewer gases, and smoke metamorphosed into the famous London fog, sometimes of a bottle-green color, sometimes pea-soup yellow." (Porter 1998, 57) Long before the Killer Smog of 1952 (see chapter 6), a week of smog enveloping London in 1873 darkened the days and led to an excess of 700 deaths.

Industrial emissions contaminated air in the vicinity of large industrial plants. The smelting of ores, for example, released quantities of sulfur that harmed vegetation, animals, and inhabitants. These problems were well known in the older industrialized regions. Freiberg in Saxony, for instance, boasted a long history of mining and smelting, and public debates about links between local industry and air pollution took place from the late 1840s. Proposed solutions came from several quarters: while farmers and foresters from the forestry academy at Tharandt tried to determine which kinds of plants best resisted the impact of smoke and sulfur, experts in hygiene and industrial technicians looked for ways to reduce emissions from the smelting plants. Their most important innovation was a chimney tall enough to disperse the emissions. The first chimney, built in 1860, was 60 meters high; it was surpassed in 1889 by a chimney 140 meters high. Notwithstanding a German proverb that referred to the "infinite ocean of the air," contemporary specialists knew very well that emissions did not disappear, but were diluted in the atmosphere and fell to the ground within a larger radius. Polluting industries profited from dilution, for it complicated the evidence of any single industrial plant's responsibility for damage to property. Neighbors and farmers made claims of compensation, but the expense of lawsuits and the disputed effects of emissions made justice difficult to achieve. Tall chimney stacks became the norm throughout much of northern Europe from the 1860s, silencing local complaints and sending the problem elsewhere. Such a solution contented representatives of the Danish hygiene movement, for example, into whose reluctant hands political authorities had placed the issue of pollution from smoke.

The decentralization of air pollution and the sheer variety of its sources also resisted control. In Great Britain, responsibility for regulating such hazards lay with local governments for most of the nineteenth century. Such a system of regulation faced notable drawbacks: local governments could not act against distant sources of pollution, they were the site of frequent conflicts of interest between officials and industry, and they embodied the basic belief that tough enforcement would drive away industrial employers. Lord Palmerston's Smoke Nuisance Abatement Act of 1853 pushed industry to implement the "best practical means" available to prevent smoke, and although prosecutions did occur while Palmerston remained Home Secretary, fines were rarely high enough to deter pollution. Besides, domestic pollution remained outside the orbit of the law. The very term "smoke abatement" speaks to the lack of political will to prevent the contamination of air.

Despite earlier national legislation, regulating air pollution on the European continent met with little more success. Napoleon Bonaparte's overarching imperial decree of 1810 constituted a precocious attempt to check industrial nuisances, and it continued to influence legislation not only in France but also in the Netherlands and postindependence Belgium. Aimed at industries that emitted "insalubrious" or "disagreeable" odors, it theoretically covered a variety of atmospheric pollutants but betrayed a distinct nineteenth-century preoccupation with odors (considered vectors of disease-causing miasma) and called explicitly for the protection only of local property owners. However, the decree allowed all residents within the vicinity of a given industry to voice complaints. French citizens became increasingly vocal during the century, gradually placing industry on the defensive with respect to residents and public authorities.

The imperial decree did not provide for sanctions or penalties of any kind. Instead, it classified industries into categories according to allowable distances from human habitation, thereby foreshadowing the twentieth-century practice of industrial zoning. Under this system a multitude of sins was allowed to continue. Moreover, France lacked an adequate corps of industrial inspectors, even after passage of a more rigorous law on industrial nuisances in 1917. That function was given to an existing corps of labor inspectors, who were overburdened and generally lacked technical knowledge or authority when confronting big industrialists. The story was much the same in Belgium and the Netherlands. Following independence in 1830, the Belgians set up a system of preventive licenses but applied it casually. A Dutch nuisance act of 1875 placed considerable power to license industry in the hands of municipalities but it, too, remained ineffectual.

The Prussian legal system similarly deterred individuals from resisting industrial contamination. Of the German states, Prussia possessed the most important industrial zones, including the Ruhr Basin, Silesia, and the region near Leipzig. In 1845 the Prussian government put into force a "trade regulation" (*Gewerbeordnung*), a law carried over by the German Reich in 1871. It stipulated that every entrepreneur who wanted to establish a factory that might harm the local environment had to request permission from the state authorities and publicize his intentions. Citizens could consult the plans and make objections, and authorities were obliged to take these objections into account. They could refuse permission or place specific obligations upon the entrepreneur; their decisions could not, however, be questioned through judicial review.

As in France, the Prussian regulation functioned less to protect citizens against pollution than to assure industry's legal security and to protect private property. If a neighbor could prove that an industrial plant had damaged his property, he might obtain compensation from the polluting firm. Most complainants were farmers owning fields in the vicinity of industry and suffering from crop damage, or fishers who lost their livelihoods due to polluted rivers. Cases were more often lost than won, however, and some courts imposed sanctions only on companies that "consciously" polluted their environments. Moreover, it was extremely difficult to establish precise links between specific plants and specific damages. At any rate, once a plant was approved, no avenues were available to close it down or to tighten the entrepreneur's obligations. Only in the mid-twentieth century was the law rescinded.

Throughout Europe, precious few authorities wished to oppose industry in the nineteenth century. Even though technicians accumulated knowledge of filters, recycling of vapors, and much else—the expression "acid rain" was first used in 1872—they rarely acted to intervene at the source. In Germany, for example, "trade taxes" on industry flowed directly into municipal budgets, so most local authorities went no further than enforcing zoning policies. About a century after the Napoleonic decree inaugurated the practice in France, local planning authorities in Germany began to delimit urban zones that were to be shielded from industry. Whereas middle- and upper-class Germans typically inhabited the protected areas, industrial workers lived in industrialized zones and had to tolerate pollution. Reinforcing the class bias of zoning practices, a significant legal principle, called *Ortsüblichkeit* (local custom), guided the German judiciary in matters of pollution. Lawyers and judges argued that inhabitants had to accept even major pollution when they lived in an area dominated by industry. The notion of "local custom" cast pollution as a part of local geography.

URBAN WATERS AND WASTES

The hazards of coal combustion and industrial emissions escaped control for yet another reason: fetid water figured as a greater and more obvious danger in the nineteenth century. Little hard evidence linked coal smoke to human health problems, yet miasma theory—though strictly a theory of airborne pathogens emanating from decaying organic matter—implicated water because water bore the load of organic wastes and was said to "exhale" miasmas. Ultimately Robert Koch and Louis Pasteur affirmed that water had cleansing powers once germs had been eliminated from it. In 1885, the International Conferences on Hygiene (1852–1908) established the procedures for analyzing water's bacteriological content, which proved to be a turning point in the combat against the recently discovered, deadly typhoid bacillus and cholera bacterium. The full significance of these events is best grasped within the larger context of nineteenth-century urbanization, with our view focused on London and Paris.

Urban growth, one of the most notable features of nineteenth-century Europe, led to degraded living environments for increasing numbers of humans, while it reduced the number of species that could adapt to profoundly modified ecosystems. The leading metropolises of London and Paris aroused sentiments ranging from wonder and ebullient admiration to terror and convictions that social disorder would remain part of urban life. Few observers combined analysis of class difference with an understanding of environmental factors to perceive the causes of poor public health, one notable exception being the English art critic and reformer George Godwin. Most bourgeois commentators expressed fears of social disorder in the context of the horrific realities of disease, for Paris and London both experienced recurrent epidemics of cholera and typhoid. Some

20,000 Parisians died in the great cholera epidemic of 1832, to name just one. For the first two-thirds of the century public hygienists attributed these diseases to miasma, and because odors were considered to be vectors, noses led experts to the fetid rivers—the Seine and the Thames.

The plight of the Seine and the Thames stemmed from the dumping of greater quantities of refuse into less water, more of which was removed for human consumption, cleansing, and the needs of manufacturers. As with coal smoke from domestic burning, this form of water pollution was not so much a hallmark of the Industrial Revolution as an effect of larger populations doing what they had traditionally done—drink, cook, clean themselves, and produce waste, both human and nonhuman, ranging from eroded soil to the offal of slaughterhouses. Inner London had grown from nearly one million people in 1800 to 2.3 million inhabitants by 1851; smaller but more densely populated Paris housed over one million people by midcentury. The Thames received more human sewage largely due to the popularity of water closets after 1815, when houses in London were allowed to connect to sewers, and especially after 1848–1849, when the Board of Health enforced the hookups. Until late in the nineteenth century, human waste in Paris was collected through cesspools, which posed their own hygienic problems. Vast amounts of mud containing much organic garbage, including horse manure, burdened the Seine. Neither city possessed anything resembling an adequate sewer system, though Paris led in sewer construction. Still, sewers remained difficult to clean, commonly overflowed into city streets, and discharged directly into the rivers. A tidal river beyond London, the Thames sucked raw sewage back into the center of the capital twice daily at high tide, which often backed up sewer outlets. Low tide was little better along some stretches where outlets were exposed, discharging their contents onto the shore.

Long-held assumptions about the self-cleansing properties of rivers yielded slowly to a will to intervene. Public officials would have to engineer purity to assure social calm and serve the basic bodily needs of enormous populations. By 1900 nearly all major cities in northern Europe had systems to separate the distribution of clean water from the evacuation of wastewater. The London Board of Health eliminated cesspools in the late 1840s, and a Metropolis Water Act of 1852 forced water companies to move their intakes upstream and regulate their filtration and storage. Drinking water showed significant improvement by the 1870s, yet the problem of the Thames—hit by 260 tons of raw sewage per day by the late 1850s—caused the most stir in the popular press as well as debate in high places. Plans for a central drainage system were stalled through much of that decade by the uncertainties of medical science and the intransigence of London's local vestries, or parish councils, which balked at centralized authorities or systems of any kind. The reappearance of cholera in 1854 and, most notoriously, the "Great Stink" of the summer of 1858 catalyzed legislation that allowed the Metropolitan Board of Works to build a comprehensive system of intercepting sewers. The Prince of Wales inaugurated this system in 1865, but the sewage, now discharged twenty-two kilometers downstream on the ebb tide, remained untreated until funding for the precipitation of solid matter began in the late 1880s.

France achieved similar changes under the authoritarian regime of Emperor Napoleon III, whose chief architect for the renovation of Paris, Baron Georges Haussmann, secured the provision of water for Paris, first through new artesian wells, which supplied pure but inadequate supplies, and then by tapping tributaries of the Marne and Yonne rivers—using parts of Roman aqueducts in the process. Haussmann completely overhauled the system of evacuation. Paris's gleaming new sewers featured tall inclined galleries, numerous reservoirs, and most of all tremendous volume to handle the wastes of the city. Though conceived under the sway of miasma theory, the sewers contributed significantly to reducing epidemics of cholera and typhoid just as scientists were identifying the waterborne bacteria that cause these diseases. Sewage continued to flow into the Seine, now downstream at Clichy and Saint-Denis, but an engineer named A. Mille succeeded in a visionary campaign to filter nonhuman sewage through the soil. "Sewage irrigation," as it was called, had already been successfully implemented by a dozen towns in England, but the solution was deemed too expensive and technically complex for any metropolis (and London never adopted it). The city of Paris established a garden, fertilized by sewage water, at nearby Gennevilliers in 1869–1870; by 1893 two-thirds of the roughly 800 hectares were in the hands of small farmers growing vegetables for the capital. The experiment at Gennevilliers was touted as a model for symbiotic urban–rural relations—a closed circle, as we would say today, of production, consumption, evacuation, and recycling. The city of Berlin adopted the Parisian model in 1878. Rising property values in suburban Paris led to the decline of Gennevilliers in the twentieth century, yet it still produced one-tenth of the vegetables sold at Paris's wholesale food market at Les Halles in 1948.

Ironically, Baron Haussmann had not wished to see his famed sewer system degraded with human feces, but several forces combined to terminate the cesspool system and usher in the *tout à l'égout*, "everything in the sewers." For one, Parisians doubled their water usage between 1870 and 1890, the evacuation of which would overwhelm cesspools in households lacking sewer connections. Second, cholera reappeared in 1884 and 1892. Finally, property owners would not tolerate new taxes to finance a special sewer system designed to carry only human waste, and complaints of odors documented a rising aversion to human

Paris's new sewers, photographed by the pioneering photographer Nadar in the 1890s.
(Hulton-Deutsch Collection/Corbis)

Passengers traveling on the Paris, Lyon, and Mediterranean Railway undergo disinfection during the cholera outbreak of 1884. Large urns of pulverized sulphate of nitrosyl are left to burn in the belief that inhaling the vapors disinfects and kills the cholera germs. Sawdust covers the floor, then is swept up and burned after the arrival of passengers, whose luggage is fumigated. Approximately 200 people died in Marseilles during the first two weeks of the outbreak, while 40,000 of the 70,000 inhabitants of Toulon left the city. (Corbis)

waste in the city's midst. Legislation in 1894 forced households to hook up to the sewer system, and water closets enabled cesspools to gradually disappear. In the ten years after 1895, the city of Paris acquired fields at three additional downstream sites, irrigating them with the now more complex wastewater that was pumped from Clichy through the Achères aqueduct. This improved the state of the Seine, which had suffered worse pollution following the construction of Haussmann's sewer system. Paris officials met their own deadline for ceasing to dump untreated sewage into the river by closing the sewer outlet at Clichy in 1899. The first regional water treatment plant went on line in 1920, complemented today by two additional plants.

Water pollution in London and Paris had its counterpart in Scandinavia. After 1850 poor sanitary and hygienic conditions led authorities in the growing capitals of Oslo, Copenhagen, and Stockholm to look to larger European cities for experience in both measuring pollution and defining solutions. The growth of Oslo—a town of 14,000 in 1816, which became an important city of over 227,000 in 1900—altered not only water quality but also the physical character of the city's watercourses. Oslo is located at the foot of hills and was once criss-crossed by many rivers, but the latter gradually metamorphosed into sewage canals as the city grew. Odors emanating from the canals became a nuisance, and the municipality covered them in the 1870s and 1880s, creating a closed sewer system from formerly natural rivers. Disappearing underground, the canals carried urban wastes out to Oslo's inner fjord. The pollution problem moved to the shores of the city.

By the 1880s experts considered the waters of Oslo's harbor basin to be severely polluted, though not nearly as bad as the Thames. (It was impossible, they noted, to see a piece of white wood three centimeters below the surface of Thames water, even in sunshine; white wood remained visible below fifty-five centimeters in Oslo's harbor.) Yet cholera and dysentery epidemics during warm summers continued. In none of the Scandinavian capitals did adjacent seas purify themselves more successfully than did rivers such as the Thames or the Seine. Bacterial content was high, and organic matter sank to the bottom and produced hydrogen sulfide, ultimately polluting the air.

Throughout Scandinavia agricultural science influenced discussions about what to do with urban sewage. As in Paris, some experts advocated the recycling of human excrement, labeled "golden manure" or the "gold of the nation" by agricultural chemists and farmers. Physicians and city planners typically regarded sewage recycling as neither economically sustainable nor hygienic or modern. They favored technical solutions along English lines, namely water closets linked to sewer systems discharging as far out in the sea as possible. Bolstered by bacteriology, the proponents of water closets won their case by the early 1920s, at a time when commercial fertilizers were gaining popularity. The system reduced the odors and high bacterial content of urban waters, but the discharge of raw sewage led to severe eutrophication (the depletion of oxygen in water, caused by algae and bacteria that feed on pollutants), peaking in Scandinavian harbor basins in the 1960s and early 1970s.

Industrialization created a distinctly new urban phenomenon—mountains of garbage. Preindustrial societies were typically too poor to throw away objects of everyday use. Clothes, for example, were converted to cloth or sold to paper mills. By 1900 increasing affluence and new materials changed the quantity and quality of refuse. Organic materials that decomposed relatively quickly had dominated preindustrial material culture. At the turn of the twentieth century, short-lived consumer goods containing glass, metal, and ceramics became widely available, and massive garbage dumps hulked along the outskirts of cities. Local governments entertained a variety of solutions. In 1896 the city of Hamburg installed the first garbage incineration plant in Germany. Its inventors touted it as hygienic and hoped to capture the energy released through combustion. Other cities experimented with separating waste. Munich's local government collected garbage and transported it to a central collecting point, where women sorted out reusable parts. At Charlottenburg, a district of Berlin, local authorities established a sophisticated waste-sorting system around 1900, which allotted three garbage cans to each household—one was for sweepings and ashes, one for food scraps and other organic material, and one for reusable or theoretically recyclable objects such as bottles, paper, and rags. A private company was responsible for garbage removal and processing. Organic material, for example, was destined as pig feed. Both incineration and waste sorting proved to be very expensive, however, and municipalities gave up waste-sorting systems until much later in the twentieth century. The Charlottenburg company collapsed in 1912; in all of Germany there were only seven garbage incineration plants by 1914.

INDUSTRY'S RIVERS AND FORESTS

The environment of Europe's largest cities received the most attention due not only to the degree of pollution there but also to the world status of these cities. Political leaders increasingly perceived malodorous waters, the scourge of epidemics, and mountains of waste as sources of international humiliation. The realms of science and technology then picked up the specific issues that had originated in politics and ideology. To continue with the theme of water, nineteenth-century industrialists, engineers, and hygienists irrevocably altered the European relationship to it in other ways as well. The great thirst of modern industry meant the consumption, with some recycling, of water in the thousands of liters per ton of coal, coke, and pig iron. Commercial interests demanded deep, straight shipping channels; steam technology and human muscle created the Manchester ship canal out of the lower Mersey and Irwell rivers, canalized the Vilaine to aid navigation at Rennes (France), and transformed much of the 1,250-kilometer-long Rhine. Farmers called for the reclamation of flood plains, altering regional hydrological cycles. Industry harnessed fast-flowing rivers such as the Rhône and the High and Upper Rhine for hydroelectricity, a form of energy whose developers promised would be renewable and nonpolluting. The dams' devastating effects on migratory fish and downstream alluvial deposits, formerly relied upon by farmers, would be felt only after scores of dams were built.

Some rivers became industrial dumping grounds for increasingly toxic effluent. Agricultural runoff burdened waterways of all sizes with too many nutrients for the dissolved oxygen to break down, leading to eutrophication. Such was the case in the heavily polluted Veenkoloniën in the northern province of Groningen (the Netherlands), where agriculture and industry together polluted an entire canal system with organic albumin from potato starch factories. Local industry was under no concerted pressure to clean up the biologically dead, stinking canals, until 1969. Likewise, the organic waste products of sugar refineries caused eutrophic conditions in German rivers. One of the first German novels inspired by environmental protection, *Pfister's Mill*, written by Wilhelm Raabe in 1884, addressed the consequences of river pollution caused by a sugar refinery.

Like water, wood became an industrial product in the nineteenth century. The Industrial Revolution modified the qualities, but did not lower the quantities, of wood demanded in commercial markets. Firewood and hardwood for sailing ships gave way to massive quantities of softwood for buildings and the transportation infrastructure. The British navy did not, however, lessen its oak requirements until the Battle of Hampton Roads (1862) during the American Civil War demonstrated the superiority of ironclad ships and brought to an end the age of the seventy-four-gun ship requiring 2,000 large oak trees. In any case, empire and free trade continued to supply Britain with wood from North America, South Africa, and Ireland; by 1914, fully 93 percent of wood products sold in England were imported. The availability of so much ghost acreage partially explains why Britain turned relatively late to the techniques of scientific forestry.

Before modern industry emerged, German treatises on forestry provided a scientific basis for the creation of the "industrial" forest, and German forestry became the standard in continental Europe. Eighteenth-century theorists conceptualized forests as sources of timber. Overall, the desirable forest became a space devoted to densely planted trees, especially fast-growing softwoods, uniform in both species and age. Such a forest permitted easy calculations of the volume of wood one could extract, and it was managed on a rotational basis, with annual clear-cuts not so large as to threaten the long-term viability of the whole forest. New growth would occur through "natural regeneration," by which trees would sprout from the seed of mature trees; this method actually required much human intervention, from the removal of competitors and understory to the eviction of livestock. The older, common method of coppicing hardwoods, that is, cutting older trees and allowing shoots to sprout from the stumps, was generally held in low esteem by foresters, for it did not yield a forest appropriate for timber production.

Modern forestry demanded a radical transformation of the natural woodland of western and central Europe—a woodland with late Paleolithic ancestry.

According to a theory of cyclical turnover, domesticated livestock continued the role once played by Europe's extinct grazers, such as aurochs and bison, by maintaining grassland into which thorny bushes like blackthorn and holly could send their underground rootstock. This thorny scrub then protected young saplings from the livestock, and a new forest, with its rough understory and open spaces in the early stages, would grow. Of course Europeans had long modified their forests through cutting and burning, but the characteristic mix of trees, bushes, and grassland had structured a resilient, species-rich ecosystem. Nineteenth-century foresters, overreacting to the higher densities of livestock in European forests due to population pressure, wished to banish all cows, sheep, and goats. Grazing rights were gradually abolished in many places, especially as fodder crops provided alternative food sources for livestock. Where scientific forestry was allowed to prevail—and it often faced contestation from pastoral communities—the result was an ecologically poorer forest, shorn of its biodiversity and made more vulnerable to disease, insect predation, and storm-felling. As the ideal river for human purposes had become a canal, the ideal forest was now a timber factory.

The Swedish experience of conflict over oak forests provides a partial counterexample to the overall trend of scientific forestry. Oak was a strategic material in Sweden as in Denmark and Norway, and the question of access to land centered on oak forests. The Swedish Crown pursued naval buildup, a project requiring oak until the later nineteenth century. Improved transportation allowed timber merchants to exploit oak forests even in remote areas of Sweden's oak belt. The Crown had attempted to regulate farmers' cutting of oak since the sixteenth century, yet farmers, too, increased pressure on oak forests in their quest for more arable land; they typically reviled oaks due to their acidic leaves and crop-damaging shade. The rural population grew rapidly in the early nineteenth century, and by then agricultural expansion jostled frequently with the royal imperative to preserve oak forests. In parliamentary discussions, representatives of the peasants' estate demanded that the Crown "liberate oak" and sell its forests to private owners. The Parliament agreed in 1818 to sell Crown forests not needed for the navy.

Then, in 1830, the Swedish Parliament decided to abolish the Crown's regulations that had long attempted to restrict cutting on common land. The Crown responded by creating specialized oak plantations under the guidance of German silviculture. Scientific calculations of volume went hand in hand with clear-cutting and the planting of forest clearings. Criticism of the Crown's forestry now came from peasants who claimed that it was difficult to obtain new growth. Private forest owners along with a German forest manager residing in Sweden, Carl Obbarius, supported peasants' assertions. Around 1850, Obbarius advanced the novel claim that the clear-cutting model threatened natural diversity. Then in the 1850s and 1860s, Scandinavian scientists pointed out the causal relationship between forest clearance and local climate change. According to their theories, weather would become colder and drier in and around vast clear-cuts. Today it is common knowledge that forests help conserve water, maintain relatively even temperatures, and regulate wind. At the time climatology played a crucial role in strengthening the case against clear-cutting in Sweden, Norway, and Finland. To the satisfaction of the Swedish Parliament, the opponents of clear-cutting signaled the poor harvests in the northern parts of Sweden and thus the connection between climatic deterioration and deforestation. Extensive debates over proper silviculture in Sweden vindicated an older model based on selective felling. This method better mimicked the regrowth that takes place in a natural forest, and it competed less with land clearance for agriculture.

Outside Sweden, however, forestry clearly separated pastures and arable land from forests. It also changed the relationship between mountains and plains, in conjunction with geology and engineering. French engineers had studied hydrographic basins since the eighteenth century, taking into account the effects of alpine deforestation in accelerating erosion and elevating riverbeds. They saw mountains as prone to disaster, continually threatening the plains with floods. And flooding mattered more in an age when plains hosted the most productive agriculture, industry, great cities, and transportation networks. In a purely technocratic spirit, the alpine states of Europe restricted communal usufruct and replanted mountain forests, often in response to severe flooding. Both French and Swiss foresters, for example, cast alpine deforestation as national calamities. Laws aiming broadly at reforestation and extending the state's control over alpine areas were on the books by 1860 in France (legislation revised in 1882 in the name of alpine restoration at state expense); and by 1876 in Switzerland, where the highland-lowland link appeared to justify this rare instance of federal intervention. In Austria, likewise, a Forest Act of 1852 removed forests from private and communal management and placed them under state supervision. Because alpine forests had a "special" role to play in protecting the plains, they were all the more subject to the rigors of scientific forestry.

The forest transition of the nineteenth century occurred alongside, and in some measure because of, the demographic transition. This shift from net deforestation to net reforestation occurred in all major European states, at somewhat different moments but largely because of the same factors. Rationalized forestry can claim some credit for this key ecological change of the modern era, but the depopulation of rural areas, and consequent natural reforestation, is undoubtedly a greater cause. Technical and commercial changes also allowed states like

Switzerland to import more wood on its rail network. All of western and northern Europe today boasts a higher forest cover than was the case in the early nineteenth century. The health of those forests will be explored in the final chapter.

EARLY EFFORTS TO CONSERVE AND PRESERVE: THE ROLES OF LEISURE AND SCIENCE

Given the deterioration of land, water, and air due to industrialization, and the rationalization of entire ecosystems, one might wonder how nineteenth-century Europeans appreciated areas not so heavily marked by industry and other manipulations. England of the eighteenth and nineteenth centuries illustrates how elite, followed by middle-class, culture guided perceptions and emerging tastes for "landscape"—a term borrowed from Dutch art in the sixteenth century and strongly connoting human modification and occupation of a given area. By the early to mid-eighteenth century, English aristocrats had begun to reject the formal, French-style garden in favor of mountains and forests. In pursuit of this new aesthetic, estate owners planted forests on their properties where there had been none. With poets and painters providing the images, the estate owners introduced dozens of exotic trees and shrubs (some 445 in the eighteenth century alone). They adorned their estates with fake temples, "ruins," curving lakes, and much else that was artificial. This trend ultimately led in the 1890s to that most artificial of landscapes, the golf course, initially greens that included some wooded areas but tending over time to the familiar grass monoculture. A different taste led to the "discovery" of the Scottish Highlands, whose intrepid English visitors of the late eighteenth and early nineteenth centuries wrote travel accounts in the Romantic vocabulary describing what they deemed to be picturesque or sublime. By the close of the nineteenth century, middle-class professionals were forming mountaineering clubs. The Highlands, and indeed the Alps over on the continent, attracted this social segment for the physical challenges provided by climbing. In Ireland, by contrast, there was little celebration of nature in the same Romantic terms; there, "landscape" and "scenery," along with estate forestry for that matter, evoked the mindset of elite Anglo-Irish landowners.

The opening up of the English and Scottish countryside to urbanites abetted landscape preservation in its early phase. More people could vacation in the Cairngorms, for example, after the opening of rail service to Aviemore in the 1860s. Walking and cycling became preferred late-nineteenth-century ways to appreciate countryside created by high farming (see above)—the pleasing "pocket-sized fields trimmed with ineluctable hedgerows." (Evans 1992, 28) Though this landscape was equally modified by human action, it was far different from the arranged aristocratic estate, as were the wilder places in England and Scotland. Images of threatened countryside gave rise to the British preservationist movement; simultaneously, walkers and hikers organized around the "right to roam" over uncultivated land in a battle for access that the British government has not yet fully resolved.

From the 1860s a perception grew in upper-middle-class circles that the city was encroaching dangerously on that picturesque countryside. In fact industry was concentrating more in cities, leaving a more rural (that is, deindustrialized) countryside, a phenomenon also taking place in France, yet cities continued to push out their boundaries. The Commons Preservation Society, founded by urban reformers in 1865 to secure and maintain access to open land, succeeded in having enclosures forbidden around London. No part of rural England could escape urban demands for food and other resources, however, as the case of Lake Thirlmere famously illustrates. The search for clean water had led city officials in Manchester to the Lake District, where in the 1870s they eyed the clear waters of Lake Thirlmere, 154 kilometers from the city. The ensuing battle between the preservationists, led by Octavia Hill and John Ruskin, and the water developers, who wished to dam the lake, "pitted the need of urban industrialism for additional resources against the need of nature lovers to retain unspoiled the rare beauty of the lakes, the most poetic of landscapes." (Winter 1999, 175) Although preservationists lost the fight for Lake Thirlmere in 1879, they emerged from it with a movement intact and a rhetoric that spoke of sacred spaces and national treasures. Many of the militants in the battle for Lake Thirlmere went on to form Britain's National Trust for Places of Historic Interest or Natural Beauty in 1895, a private entity whose property was made inalienable by Parliament in 1907. Significantly ahead of its time, the Society for the Promotion of Nature Reserves was created in 1912 for the purpose of identifying endangered habitats that well-funded organizations, such as the National Trust, might purchase.

While preservationism was beginning to mobilize prominent public figures in Britain, the "nature and homeland protection" movement emerged at around the same time in Germany. The leading figure in early German conservation was Hugo Conwentz, a botanist, bureaucrat, and prolific writer and lecturer who contributed to the definition of "natural monuments," encompassing specific natural treasures within a cultural landscape. The crusade to protect natural monuments broadly influenced European conservation, and it stands apart from the contemporary American concern for wilderness preservation. Discourse around nature and homeland protection spawned key institutions, including the private German League for the Protection of the Homeland (1904), instigated by composer and musicologist Ernst Rudorff, and the Governmental Center for the Conservation of Natural Monuments, established in Prussia two years later. Decidedly anti-urban and anti-industrial, conservation's bourgeois constituency expressed unease over many phenomena, including deforestation, the general acceleration of modern life, and urban popular culture. Their facile remedy for such ills was to advocate a return to "traditional" rural life, but concretely they fought against hydroelectric power plants and other industries that spoiled the visual qualities of inherited landscapes. They advocated a new architectural style to mesh with rural landscapes, a "homeland protection style" inspired by rural dwellings.

Perceived as "habitats," some landscapes became valued for the rare or endangered species they hosted; in Britain, one class of animals—birds—received by far the most attention. Despite the habitat afforded by hedgerows, raptors faced extinction by gamekeepers whose brief was to protect grouse for hunters. Many species of birds had become endangered by the trade in plumage for women's hats. The female leadership of the movement formed a Society for the Protection of Birds in 1889 (called the "Royal Society" after 1891, and today the largest conservation organization in Britain). This society can largely be credited with a series of laws which, though weak, and ignoring game birds, did begin to change consciousness about the wanton killing of many wild birds and the collecting of eggs. Continental European societies had also begun to protect certain species. In France, for example, the Grammont Law of 1850 punished with fines and imprisonment anyone caught maltreating a domestic animal in public—testimony to a bourgeois "civilizing mission" enacted in a still highly rural country. As in Britain, upper-class Dutch women promoted kinder treatment toward animals; their efforts helped bring about the Useful Animal Act of 1880, a law protecting only animals deemed serviceable in agriculture or forestry. By the end of the century, scientists in the Netherlands had marshaled ecological arguments to show the limitations of "usefulness" as defined in 1880. In nature all animals were useful.

As was happening in North America, European scientists—biologists, naturalists, ecologists—were appropriating animal protection issues from upper-class, most often female, volunteers. Amateur natural history societies arose in Europe in the eighteenth century, and naturalists stood at the helm of the emerging discipline of ecology prior to its institutionalization in the early twentieth century. Provincial natural history societies flowered in nineteenth-century France. Imbued with Jean-Baptiste Lamarck's evolutionary ideas, they worked with ecological notions—briefly, the relationships between species and their environments. Shortly after midcentury, Charles Darwin's groundbreaking treatise on evolution, *The Origin of Species*, not only called into question the Frenchman's ideas of inheritance but also redefined the essential relationship between environments and species: Darwin cast nature as a set of evolving places constantly up for grabs by species competing with one another. A few years later, German scientist Ernst Haeckel decided that a new name was in order to denote the developing field concerned with those relationships, and the term "Oecologie" was born.

Whether they were naturalists in the Lamarckian mode or Darwinian ecologists, scientists often contributed to the cause of nature protection, but other members of the professional middle class did likewise. In Third-Republic France (1870–1940), a preservationist movement rested on an innovation from 1830, the Inspectorate of Historical Monuments, which had gone far to institutionalize a French notion of "heritage." By the end of the century, heritage became joined with the discovery of nature in a patriotic package that bore resemblance to the English and German variants noted above. French tourist lobbies such as the Alpine Club and the Touring Club militated in favor of tree planting and alpine restoration. The pedagogical flavor of French preservationism at the turn of the century combined the popularity of Arbor Day (the Fête de l'Arbre) with the schooling of children in the tremendous natural diversity of their country. Voted in the name of patriotic attachment to "native soil," the 1906 law on the Protection of Natural Sites and Monuments was the first of its kind in Europe. The text revealed a century's legacy of artistic taste: sites worthy of protection would be defined according to their picturesque qualities. Well beyond France, the subjectivity of such criteria long hounded the case for protecting landscape.

CONSERVATION IN THE ERA OF WORLD WAR AND ECONOMIC CRISIS

The two world wars of the twentieth century both accelerated and retarded the major environmental trends that had been set by the Industrial Revolution. Entire sectors of the consumer economy were necessarily placed on hold, but the skewed production for war caused the belligerent European powers to increase their ghost acreage in search of industrial resources as well as food supplies.[2] On the one hand efforts at pollution control and nature conservation nearly halted as national priorities shifted dramatically; Britons even celebrated smoky air as a defense against aerial bombardment during World War I. On the other hand, each war would in turn stimulate new thinking about both pollution and the use of resources. Applying such a "before" and "after" framework to each war, however, obscures the environmental devastation that twentieth-century warfare caused in Europe.

The pummeled landscape of northern and northeastern France during the first world war, and the bombed and burned cores of German cities during the second one, figure among the worst historical examples of wartime destruction. In World War I, much of French industry, the Pas-de-Calais coalfields, the country's most productive agricultural zone, and the vibrant towns of Flanders all fell behind German lines for the duration. After four years of being ripped apart by heavy artillery and trench construction, soils, fields, and some 200,000 hectares of forest lay in ruins. France faced alarming wood deficits due to the devastation and the war's appetite for wood. In this as in the German case after World War II, tremendous national will and international aid were funneled into postwar restoration projects. Even the most highly devastated "red zones" of northern France had undergone soil restoration and reseeding by the end of the 1920s. Notwithstanding such an achievement, at the end of the twentieth century the occasional discovery of an unexploded shell in a French field served as a grisly reminder of the long half-life of industrialized combat.

As World War I caused entire forests to disappear in France, massive felling for the war effort sparked official anxiety in Great Britain as well. In a country that was importing 90 percent of its timber and forest products by 1900, the cutting of half-a-million acres of forest during the war was anything but "sustainable." Britain came late to official forestry, creating the Forestry Commission in 1919, with a clear mandate to expand the forested area of the country to produce domestic timber. Some politicians linked forestry with hopes for rural development and the resettlement of demobilized soldiers; the Forestry Commission itself prioritized expert management yet quickly encountered resistance from private estate owners, who possessed nearly all British woodlands. For the latter, the prospect of uniform, softwood forests posed as an alien danger to a cherished ideal of a less orderly, more diverse countryside. Preservationists and the tourist lobby, too, held some influence over the Forestry Commission, which was providing recreational facilities within its Forest Parks by the late 1930s.

Conservation during the interwar years remained a socially genteel, minority pursuit in western Europe. In 1926 architect and town planner Patrick Abercrombie founded the Council for the Preservation of Rural England, with analogous councils for Scotland and Wales established in 1927 and 1928. They were arguably the first associations of their kind whose members were determined to "act locally," as a later generation would put it. Until real limitations on development were implemented after 1947, however, they largely were unable to prevent the loss of more than 25,000 hectares per year, as the British built four million houses between 1920 and 1940. The location of many new residences required owning a private automobile, a growing phenomenon addressed in the Restriction of Ribbon Development Act of 1935 ("ribbon development" referred to continuous housing along roadways that blurred the boundaries between towns and also between town and countryside). Over three million cars plied the roadways of Britain by 1939. This technological centerpiece of the twentieth century was already affecting not only development but also urban air quality, vacation patterns, and family economies, though its full environmental and social impact would not be measured until the decades following World War II.

The interwar period in Germany saw broadened environmental consciousness despite most citizens' preoccupation with a crisis-ridden economy. The constitution of the Weimar Republic (1918–1933) stipulated that public authorities must protect landscape and nature. Article 150 placed the protection of artistic, historical, and natural "monuments" under the care of Germany's individual states, but as the Weimar Republic experienced frequent political crises, no important innovations could be realized. Fortuitously, the political events of 1923 showed the inhabitants of the Ruhr Basin what life might be like without the hazards of industrial pollution. When French troops occupied the region, German politicians and trade unions organized a general strike that lasted from spring until autumn, effectively shutting down all coal mines, coking plants, and steel mills. The air became fresh as emissions ceased. For the first time in living memory vegetables flourished in gardens, leaves remained green right into autumn (whereas they normally turned brown in summer), and fruits did not have their usual layer of soot and tar. Despite the general relief felt when production restarted (the economic situation had become intolerable), the idyllic summer of 1923 catalyzed industrial experts to forward proposals that would alleviate the Ruhr's intense pollution. For example, they advocated the development of residential heating systems that would use waste heat from industrial plants. Unfortunately, large firms balked at these ideas. The period of the Weimar Republic saw much expertise turned to environmental problems, but economic and political instability prevented lasting improvements.

The Third Reich (1933–1945) began with ambitious goals for nature protection. From the outset the new government issued laws that punished cruelty toward animals and regulated hunting. In 1935 the Nazis imposed the Imperial Nature Protection Law (*Reichsnaturschutzgesetz*), which set up nature reserves and protected areas and mandated a specialized conservation administration that had to be consulted in case of potential harm to landscape wrought by construction or industry. In theory the conservation administration had the power to thwart harmful practices, but as was the case with many aspects of Nazi governance, nature protection formed an isolated goal that did not necessarily conform to other policies. It remained in conflict with policies of economic reconstruction, autarky, and militaristic expansionism. With Germany's invasion of Poland in 1939, the wartime economy forced increases in industrial and agricultural production, and the immediate impacts of military actions contributed to the deterioration of environmental conditions far beyond the borders of the Reich.

Moreover, environmental protection in the Nazi period was not "innocent." On the one hand, conservationists of the 1920s had adopted *völkisch* rhetoric purporting that landscape and soil fundamentally molded the character of a people. On the other hand, they claimed that the German landscape demonstrated the racial superiority of its inhabitants because they had allegedly forged the quintessential rural, and most aesthetically pleasing, landscape out of wilderness. Through its constant battle against nature, Germany's agricultural population had proved its special ability to win the struggle for survival of the fittest. Supposing a deterioration of the German people since industrialization, *völkisch* conservationists advocated a kind of "ruralization" of the populace. In this respect, their ideas coincided with those of Heinrich Himmler, leader of the SS and responsible for the reorganization of Poland and other conquered territories in eastern Europe. Conservationists, landscape architects, and planning experts on Himmler's staff developed the "General Plan for the East," whose goal was to create a kind of ideal "German" landscape in the occupied regions. It was a strange mix containing elements of modern landscape protection (taking account of local hydrology, for instance), agrarian romanticism (exemplified by the prescribed "rural" style of houses and villages), and industrial development. Above all, the plan was inspired by brutal racism—eliminating the existing population was a prerequisite for its realization. After World War II, the proponents of the General Plan vehemently denied these links to inhumane Nazi policies. It is worth noting that a majority from the old homeland protection movement backed these plans, hoping to improve their political influence in the Third Reich.

Decades before the human and environmental horrors of World War II, many of the leading advocates of nature protection in northern Europe, including Hugo Conwentz and Dutch biologist Frederik Willem van Eeden, had adopted the concept of "natural monuments." The Dutch Naturalist Society proposed a definition in 1905: "an otherwise disappearing part of nature, which ought to be saved in a civilized nation for future generations, because of its value for nature study, for enjoyment, for science, and for poetry." (Windt 1999, 238) The phrase "an otherwise disappearing part of nature" highlights what was distinctive about the European preservation movement in its initial phase. It rested on the perception that much of nature—however cultivated or modified by other human activities—*was* disappearing, and civilized humanity had to safeguard portions of it where possible. Whether scientifically worthy or beautiful by artistic convention, those islands of nature appeared few and far between as modern industry and urban infrastructures left heavy, indelible marks.

Notes

1. Even North American ghost acreage (imports that increase an area's available acreage) could not supply Britain with sufficient iron, despite the fact that by the 1750s a band of counties centering on northern New Jersey and east-central Pennsylvania were producing more iron from charcoal than were the British Isles.

2. According to John McNeill, over six million hectares of North American grasslands were plowed up in order to meet the European demand for wheat during World War I. See McNeill, *Something New under the Sun: An Environmental History of the Twentieth Century,* 346.

6

CONSUMER REVOLUTION AND GREEN REACTION
Economics, Ecology, and Environmentalism since World War II

As late as the mid-twentieth century, Europeans did not yet use the concept of "environment" in its most popular, present-day meaning, informed by ecology: the multitude of organic and inorganic elements that surrounded, shaped, and were increasingly threatened by human civilization. Ecological science had for some decades applied the concept to *other* species, and the social sciences spoke of political, cultural, or social "environments." French, German, and Swedish speakers likewise used the words *milieu*, *Umwelt*, and *miljö*, respectively, in these contexts. Europeans were on the brink, however, of a new set of experiences with modified nature, born of unprecedented affluence that followed the most devastating war in history. Further revolutions in energy use, industry, agriculture, and urbanization, as well as the birth of what we call "consumer society," would also give rise to a heightened consciousness of the toll that growing European gross domestic products were taking on the limited land and waters of westernmost Eurasia and, indeed, many other regions of the world. Consciousness of scarcities, degraded soils, polluted rivers, and much else was felt and acted upon in many different ways across European society, ways that included the production of theoretical "doomsday" scenarios, activism through nongovernmental organizations, and action within the existing structures of politics. This chapter will outline the story of those revolutions in ecology and economy but can only begin to narrate the story of environmental awareness.

POSTWAR GROWTH AND DEGRADATION

Millions of European experienced degraded environments during World War II: destroyed infrastructures, inferior food and less caloric intake, and plummeting standards of water purity were just a few facts of daily life throughout much of the war. Where industry had been bombed into quiescence, reduced or nonexistent emissions favored local lungs as well as vegetation for a time, but only after the bombings themselves had caused tremendous releases of gaseous pollutants. Firebombed cities became sites of gruesome destruction and contamination. Scientists in the public service devoted themselves less to curbing air pollution than to gathering intelligence and developing radar and atomic weapons.

Long before the fighting ceased (on May 8, 1945, in Europe), leaders understood that their populations would come out of the war with great expectations for peace and stability. One line of thinking led to widening the protection afforded by European welfare states, a means of stabilizing the postwar democracies in the throes of recovery. A second line applied the wartime values of strategizing and planning to domestic policy. Of consequence to the environment, Britain's powerful Town and Country Planning Act of 1947 invested county councils with the authority to give or withhold permission for new developments or changes in land use. The law included such possibilities as creating green belts around cities to prevent sprawl; green belts around British cities accounted for 1.6 million hectares by 1984. Local interests, however, eventually collided with other visionary measures, namely the will to create and manage national parks (see below).

The larger goal of planning, however, targeted not nature conservation but rather economic growth and technological advancement. Whether invaded and occupied by the military forces of Nazi Germany (France, Belgium, the Netherlands, Denmark, and Norway) or forced to spend the national wealth defeating the aggressor (Great Britain), northern European countries faced economies in shambles in 1945. For French leaders and citizens, the shocking defeat of 1940 would resonate for decades to come, and the slow-growth, still highly agricultural economy of the earlier twentieth century had tragically proven its deficiencies. Combining elements of free-market and planned economies, technocrats led by Jean Monnet put together voluntary four-year plans affecting all economic sectors. The French government of the new Fourth Republic also nationalized coal, gas, and electricity, in addition to nationalizing civil aviation, deposit banks, and the Renault car manufacturer. Similarly, Clement Atlee's Labor government in Britain placed the Bank of England, energy utilities, and transportation under public ownership. Both governments also approached planning from the perspective of regional development: French planners sought to restore balance to an economy tilted in every way toward Paris and its suburbs, whereas British politicians looked at specific depressed areas as they passed the Development Areas Act in 1945.

The firebombing of Dresden by the Allies in February 1945 brought total environmental destruction to the city's residents. It remains one of the most controversial examples of aerial bombardment during World War II. (Library of Congress)

The fundamental change in the European energy regime—from the primacy of coal to the primacy of oil—affected the location and nature of industry more than did government planning, however. Northern Europeans contributed relatively little to the birth of the twentieth-century oil economy: Imperial Russia and the United States pioneered exploration and production in the nineteenth century, and the only oil field in northern Europe worthy of note—the North Sea—did not become important until after OPEC restricted its oil production beginning in 1973. As a predominantly oil-poor region, northern Europe looked to import petroleum, which promised greater efficiency and more numerous applications compared with coal. Already the dominant fuel in transportation prior to World War II, oil nudged out coal in industry, electrical generation, and domestic heating as new oil fields opened and complex infrastructures for refining and transporting petroleum kept apace after 1945. Following the war, oil prices fell for nearly three decades. Industrial regions associated with the old coalfields, such as the north and east of France, lost their advantages; oil refineries and petrochemical plants concentrated aggressively around port cities such as Le Havre and Rotterdam. Not only did Dutch officials rebuild much of the latter port to accommodate oil tankers and the refineries of a half-dozen major producers, but they also attracted some thirty chemical companies to the surrounding region, baptizing the urban-industrial monolith "Randstad Holland." In his eco-biography of the Rhine river, Mark Cioc describes a segment of the Nieuwe Waterweg (which connects the Delta Rhine to Rotterdam) as a "seamless web of oil tankers, petroleum refineries, petrochemical plants, storage facilities, and port terminals." (Cioc 2002, 135)

If Randstad Holland epitomized the landscape of oil transport and refining, the rest of northern Europe absorbed, in one way or another, the environmental effects of oil-based manufacturing and transportation. Coal had largely confined industry to the hinterlands of the coalfields themselves, whereas the relative cheapness of oil allowed industry, and thus pollution, to decentralize. Oil meant pipelines and, as the major fuel behind electrical generation, high-tension cables, power plants swallowing up to 200 hectares of land apiece, and electrified railways. Importation of crude oil led to bulging tankers coming into port and occasionally spilling their contents in European coastal waters. French leaders somewhat lessened their country's reliance on imported oil by launching a massive nuclear program (the Plan Messmer) in 1974, yet this exception served to underline the galloping consumption of oil occurring nearly everywhere on the European continent.

The tremendous growth in energy consumption was one of the midwives of postwar consumer society. Cheap crude oil fueled the greatest economic boom in European history, nearly thirty years of growing GDP per capita. Historic in-

Oil storage tanks at Europort. Lying at the mouth of the Rhine, Europort is one of Europe's largest ports. (Michael St. Maur Sheil/Corbis)

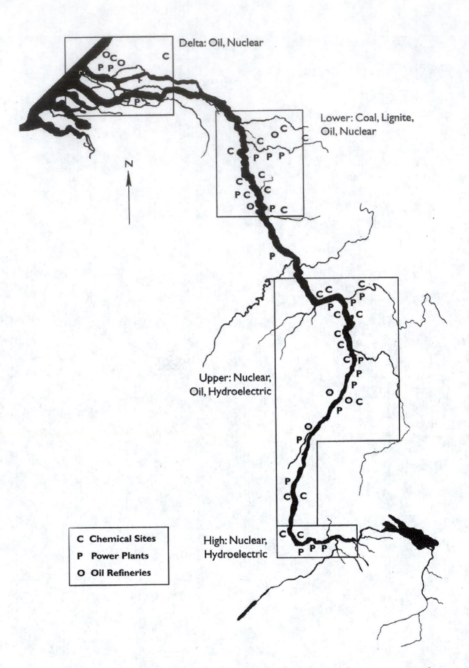

Chemical sites, power plants, and oil refineries flank today's Rhine, Europe's longest industrial river. (Mark Cioc, The Rhine: An Eco-Biography, *1815–2000, 2002)*

ternational agreements extended the role of cheap energy by triggering growth: Bretton Woods (1944) stabilized exchange rates among currencies, the General Agreement on Tariffs and Trade (1947) sought to dissolve trade barriers among its members, and the Marshall Plan (1948) supplied badly needed initial capital. As growth picked up, many workers confidently made the switch to advanced manufacturing as well as service sectors. Above all, they abandoned farming, lured by high wages and the attractive working hours and paid vacations recently won by organized labor. Those fundamental shifts often entailed moves to larger towns and cities, focal points of advertising and retail, or, put another way, focal points for the multiplication of desires. Statistics for France tell part of the story for much of western and northern Europe: in the third quarter of the century, energy consumption quintupled, the number of automobiles multiplied by fifteen, and the consumption of plastics multiplied by twenty.

Consumer society produced waste as never before. Household recycling and reuse of materials did last well into the twentieth century, in large part because of the privations of two world wars and the economic crisis of the 1930s, but a postwar "throwaway mentality" exacerbated the problem of garbage from the late 1950s. The West German "economic miracle" (1954–1974) enabled the population of West Germany to rapidly achieve the average standard of living prevailing in the United States. The widening range of everyday products now included synthetic materials that complicated recycling and reuse. Towns continued to dispose of waste in garbage dumps, but by the 1970s German officials deplored the excessive space required by the dumps, and they were concerned about the danger of toxic substances from the dumps leaching into and contaminating groundwater. Most local authorities in West Germany reintroduced waste-sorting systems and biological processing of organic materials in the 1980s. The federal government pioneered the collection and reuse of packaging material in its widely heralded "Green Point" system. In contrast to the private garbage-sorting ventures of the early twentieth century (see chapter 5), the federal government sought environmental protection rather than a profit from garbage. Nevertheless, the effectiveness of both local and federal schemes remains disputed. Some experts doubt, for instance, whether the cleaning and reuse of glass bottles is environmentally friendlier than the onetime use of aluminum- or plastic-coated cardboard boxes, or "Tetra Paks."

The East Germans had established a sophisticated system of waste sorting, called the "SeRo-System," much earlier. An extreme scarcity of resources and the communist government's goal of economic self-sufficiency lay behind its adoption. Rare materials like paper and aluminum were gathered in a very effective manner. The system managed to collect around 40 percent of domestic and

The "Killer Smog" of 1952, seen from Blackfriars, London. (Getty Images)

industrial waste, and recycled materials accounted for 13 percent of industrial raw material in the German Democratic Republic. This system functioned until German reunification in 1990. Environmental protection itself received little attention in the Democratic Republic, where there were nearly no controls on toxic materials issuing from industrial production. In 1990 the famous Silbersee (Silver Lake) in Bitterfeld near Halle was discovered to contain tons of mercury—an extremely dangerous heritage of a nearby chemical factory.

In the energy regime of postwar Europe, incentives to employ energy-saving technologies were few. Air pollution intensified throughout the industrialized world until the end of the 1960s. Londoners, who continued to burn coal in open fires after World War II, experienced the "Killer Smog" of December 1952, a Victorian-style "fog" (see chapter 5) that caused some 4,000 excess deaths and spurred the passage of a Clean Air Act in 1956. Targeting domestic sources of pollution, the law allowed local governments to create "smokeless zones" where all sources of black smoke would be subject to regulation; these zones came to cover over 90 percent of London. By then, oil, gas, and electricity were becoming competitive with coal for heating purposes. The substitution of fuels helped dissipate the pall of smoke over London, yet more sunlight in this and other cities created more photochemical smog as greater numbers of cars spewed hydrocarbons and nitrogen oxides into the urban atmosphere.

Gluttonous use of energy and its significant environmental consequences did attract official attention, although governments within northern Europe did not react identically or simultaneously. West German legislation began as early as the 1950s and 1960s to attach greater importance to health care in the context of industrial policy. Concerning air pollution, West German authorities developed new policies based on what would later be called the "precautionary principle." Instead of regulating recurring problems on a case-by-case basis, regional authorities in the Ruhr Basin began to fix threshold values, defined in terms of human health, as standards for limiting emissions. Though it was far from holistic in approach, legislation now put human health on par with the protection of private property. The so-called technical instruction on air (*Technische Anleitung Luft*) of 1964 and the Federal Emission Law of 1974 codified these principles. For instance, they enabled the concerned authorities to require modifications of production plants, even if an original plant design had previously met with approval. Despite its limited political influence, the new Ministry of Health, created in 1962, began to outline systematic health care policy that encompassed all possible dangers to human well-being.

In West Germany, the quality of water remained critical until the turn of the 1960s, when local authorities and firms began to build sufficient water-treatment plants to decontaminate river water. In East Germany, by contrast, the quality of air, river water, and groundwater worsened until German reunification in 1990. Initially the communist government had tried to integrate ecological principles within the planned economy. As economic results did not meet early expectations, the government prioritized agricultural and industrial production: poor levels of productivity did not allow for expensive environmental technology. One of the main reasons for the poor performance of the East German economy (compared to the West) was the shortage of foreign exchange,

An open pit coal mine with a distant cloud of smoke hovering over a coke plant and steel mill in Germany's Ruhr Valley. (Time Life Pictures/Getty Images)

which prevented East Germany from importing energy, particularly oil, from Western countries. Instead, the country depended on supplies from the Soviet Union and on its own resources. The most important natural resource was brown coal mined in the area centering on Leipzig and Halle. East German brown coal had little calorific value but contained a large proportion of sulfur; sulfur dioxide, a byproduct of combustion, contributed to high levels of bronchitis and other respiratory illnesses among the inhabitants of highly industrialized regions. Finally, there was little civil society in the German Democratic Republic to counter official politics. An ecological movement did come into being in the 1980s, and although its members did not intend to overthrow the regime, they contributed decisively to its final destabilization, not least because the government proved unwilling to accept the existence of glaringly obvious problems.

In the capitalist countries, the growth of human habitat and transportation networks outpaced regulation. Automobile traffic and urban development increased as suburbs sprouted and smaller towns grew; following the American model, Britain and Germany led the way in suburbanization, while middle-class French people pioneered in acquiring second homes. With both outward growth and urban densification, cities deepened their ecological footprints. New road systems and renovated railways allowed building materials to come from further afield. As a rule, environmental problems could be localized before 1950: pollution largely afflicted the industrialized, densely populated regions like the Ruhr Basin. From the 1950s the range of pollution widened due to a more decentralized population and suburban sprawl that replaced arable land, woods, and other open spaces.

From another angle, the triumphant advance of consumer society in northern Europe created a new polluting regime that implicated individual consumers more than ever before. They polluted the ambient air when they drove their automobiles; they sullied waterways when detergents exited from millions of private households; they helped destroy the crucial layer of ozone in the upper atmosphere as they used products containing chlorofluorocarbons and other gaseous mixtures. In comparison, controlling industrial production is relatively simpler because industry is centralized and can be subjected to technical standards. It has proven far more difficult to influence the behavior of the individual in liberal western societies. Therefore, the modern environmental movement (see below) not only called upon the state but also appealed to the average citizen to adopt a less destructive lifestyle.

FLUCTUATIONS IN FISHERIES

The postwar expansion of capitalist investment and markets in a new technological setting also affected traditional activities, fishing among them. Historically the Scandinavian countries have been nations of fishers. Many people fished or could obtain fish. From the end of World War II new technologies dramatically increased the scale of fishing around the world, a scale exceeding the rate of renewal for many Scandinavian fisheries by the 1960s. Between 1970 and 1990 the world's fishing fleet doubled, and European fleets today have an overcapacity of between 40 and 60 percent. In Scandinavian waters like the Baltic Sea, overfishing has reduced the catch such that Baltic salmon, eel, and herring are threatened with extinction.

One of the traditional fisheries in Sweden and Norway since the Middle Ages has been herring. The quantity of fish caught fluctuated according to natural conditions. When the herring catch was good in Sweden it was often bad in Norway, and vice versa. Fishing technology had little influence on yields. In the early twentieth century the herring yield improved considerably along the west coast of Norway, peaking in the late 1950s at nine to ten million tons. Beginning in the same decade, however, fishing fleets equipped with sonar and seine netting depleted stocks all too efficiently. In order to recover their large investments in new technology, the fleets had to fish nearly around the clock, putting pressure on herring populations. The species was close to extinction by the early 1970s. Herring fisheries saw some improvement in the 1990s but remain tightly regulated.

One area of particular interest to the Scandinavian countries is the Barents Sea, constituting 1.4 million square kilometers of shallow water between the European subcontinent and the Arctic Basin. Large seasonal fisheries based on pelagic species, namely cod, long supported the local fishers' livelihood and were central to subsistence in the northern parts of Scandinavia. One of the world's most productive fisheries was also one of the most commercialized.

Until the late 1970s, the Barents Sea yielded generous catches. In 1980 it reached approximately 2.4 million metric tons of fish, or 3.75 percent of the global catch. Ten years later the catch was only a fraction of the 1980 catch, and the cod fisheries approached ecological crisis. Through multilateral and bilateral agreements within the coastal zones, a system of quotas had existed from the late 1970s. In the Barents Sea, however, fishing takes place in areas marked by different legal controls. For instance, in international waters fishers may fish according to an open-access rule, and during the 1990s fishing in international waters increased. The fishers found an area free of regulation between the Spitzbergen Archipelago and the island of Novaya Zemlja. In an effort to prevent extinction and stop cod fishing by Greenlanders in this zone, Norway and Greenland signed an agreement in 1991. The European Union then forbade fishing in the Barents Sea by its member states by the end of 1992. In the 1990s the dominant fishing nation in international waters in the Barents Sea was Iceland, and its fishing fleet's presence there elicited protest from both Norway and Russia. The Norwegian coastguard even cut the lines of Icelandic trawlers. In 1998 the three parties cooperated in establishing a fishing quota and signed an agreement in 1999.

Efforts to regulate fisheries in the Barents Sea have been described as an attempt to avert a "tragedy of the commons."[1] During the 1990s the fisheries' recovery seemed to vindicate the regulations. A fundamental question remains: did the crisis and its resolution depend only on human activities, or did variations in marine ecology play a role? The most popular explanation has focused on overfishing and management, with sociologists pointing at fishers' greed combined with an absence of public understanding of fisheries. Scientific advice provided to international authorities during the 1980s largely determined the quotas, however, and it now appears that the precipitous decline in fisheries of the 1980s was also due to natural fluctuations, namely variations in climate.

THE LATEST REVOLUTION IN AGRICULTURE: HYPERPRODUCTIVITY

The unprecedented scale of industrial expansion, the advent of consumer society, increased urbanization—none of this would have been possible without a new revolution in agriculture. Lest our image of postwar Europe rest with the triumph of urban-industrial culture and its aggregate environmental effects, we must turn to the vision of an endless, brownish-yellow wheat field in central France, or the brighter yellow of a rape field in England. Postwar agriculture continued to account for far more acreage than did cities, roads, and industry combined, and its social and environmental consequences were on par with those linked to cities and manufacturing. As seen in the previous chapter, farming had modernized and specialized in most of northern Europe long before 1950; agribusiness itself, implying large fields, tractors, chemical fertilizers, and biocides, began reshaping the English countryside after World War I. After 1945 political decisions steered agriculture decisively toward hyperproductivity. Programs for food self-sufficiency were launched in the name of national security; in this as in other areas, memories of shortages from two world wars played a significant, if not determining, role.

Furthering decisions made at the national level, the European Economic Community (nicknamed the Common Market) set about formulating what became the Common Agricultural Policy (CAP) in the late 1950s. Devised to accelerate productivity in the interests of both farmers and consumers, the CAP provided direct price supports for most agricultural products. The Common Market met and surpassed its goals, and despite the imposition of quotas designed to limit production, food surpluses became a fact of life. Just as the European Union (EU) began to scale back grain subsidies in the early 1990s, the twelve member states were producing grain surpluses of 25 percent. The CAP budget accounted for a whopping 44 percent of the European Union's expenditure in 2000. Automatic, direct subsidies went hand in hand with the technical advances that allowed the land to yield ever more food for humans.

The Green Revolution is not usually associated with Europe because its effects were less revolutionary there than, say, in Mexico or India. Yet Europe experienced a Green Revolution early on, and it played into political decision-making. New varieties of wheat, maize (known as corn in the United States), and rice transferred biomass to the seed-packed head of the plant, causing astonishing jumps in productivity per hectare—as long as farmers applied specific packages of fertilizers and biocides to each crop. These costly inputs required economies of scale; thus the popularity of hybrid wheat and maize helped drive the trend toward larger fields and the requisite tractors and combine harvesters.

Beyond the cultivation of hybrid cereal crops, intensive agriculture took many forms, from state-of-the-art greenhouses in the Netherlands to industrial pig and poultry farming in Brittany. Marshall Plan funds helped with early purchases of tractors and other equipment, yet indebtedness soon became a constant in the lives of many farming families, and the CAP benefited large producers more than small ones. Few observers (and certainly not government officials) wrung their hands over the demise of a familiar rural world in the early postwar decades, however; the gains in productivity were far too spectacular. Britain achieved its self-sufficiency in grain by 1986. Today that country of modest size boasts the largest number of sheep, the third-largest grain crop, and the greatest proportion of large farms in the European Union. France, whose land mass is slightly smaller than Texas, is the world's second-largest food exporter after the United States.

The destruction wrought by pesticides on wildlife and elaborate food webs, the pollution and eutrophication of waterways from nitrogen-laden agricultural runoff, the loss of crop diversity—such effects were felt everywhere intensification made its mark, and Europe was no exception. In Britain the new synthetic pesticides began to take their toll on bird, badger, and otter populations from the early 1950s. Despite the evidence publicized by conservation organizations, control of the worst offenders happened in fits and starts. The British government regulated organochlorines (chlorine-containing pesticides) from 1961, but a mandatory ban on all uses of aldrin and dieldrin appeared only in 1989. West Germany prohibited the use of DDT in 1971, and the European Community began to enact legislation pertaining to pesticides in 1979. In the context of the Sixth Environmental Action Programme, the European Commission adopted new strategies for pesticide control in June 2002; these included plans to monitor distribution and use more closely, and to reward low-input and pesticide-free farming with financial incentives. A year later the EU began a phaseout of some 320 pesticides that had not met rigorous standards. Such action should be seen within the larger shifts in the EU's stance toward agriculture, detailed below.

Agricultural runoff could pose more immediate threats to human health by contaminating water supplies with phosphates and nitrates. Regulation has been sporadic and piecemeal, though occasionally swift and effective, as in the case of Lake Constance and Stuttgart. Intensive fruit and vegetable production dominates the region to the north of Lake Constance, a lake shared by Germany, Switzerland, and Austria. In the 1960s fertilizers drained into the groundwater and altered the biological composition of the lake. Because Lake Constance served as the main water supply for the densely populated agglomeration around Stuttgart, the drinking water of millions of inhabitants was in danger. As a consequence, regional authorities installed a sophisticated system of water treatment plants. In 1979 the water of Lake Constance contained more than 50 micrograms of phosphate per liter, but by the end of the century the proportion had declined to 14 micrograms—still well above a German drinking water regulation of 1999 that fixed the threshold value for phosphate at 6.7 micrograms per liter.

SUSTAINABLE SUBSISTENCE: THE SAAMI IN POSTWAR EUROPE

This latest agricultural revolution brought more segments of rural northern Europe within the orbits of capitalist production than ever before. A look at the recent evolution of the northern Scandinavian Saami culture shows how even the last remaining group of subsistence producers in northern Europe faced the pressures of consumer society and capitalist production at the expense of ecological stability.[2]

The original Saami culture was anchored in the right to use land and water in the Saami homeland, which includes parts of Norway, Sweden, Finland, and Russia. The Saami have perpetual usufructuary rights to land on which they herd reindeer, hunt, fish, and fell timber. Their traditional subsistence depended on mobility, and the Saami people grew accustomed to long-term swings in the availability of one or more critical resources. With knowledge of their animals, terrains, and each other, groups of pastoralists divided their large areas into pastoral ranges, inhabiting a basic unit called the *sii'da* and moving their animals between coastal summer and inland winter pastures. The Saami recognized ecological uncertainty and could think in modern terms of sustainable development. When resources fell short, reindeer starved and the herds were depleted. Periodic crises thus helped regulate the pastoral system.

Ways of handling resource scarcity have changed in tandem with changes in Saami and Scandinavian society. Many herders have chosen to use snowmobiles, for example, in order to follow the herds more closely. In turn they have increased the size of their herds in order to pay for their new equipment. Overcrowding now burdens the winter pastures, causing overgrazing and soil erosion. Authorities point to both problems in locating the major ecosystemic changes occurring in Norway today. In particular, lichen pastures have suffered much in the past ten years, with the necessary area covered by lichen gradually decreasing. Traditional transhumance between coast and inland secured high-quality pasture; larger numbers of reindeer and the building of fences now impede free movement between pastures. In Finland the Saami people no longer move their herds, and lichen pastures have disappeared.

Saami herder with reindeer, Norway, 2002. (William Findlay/Corbis)

One might ask why this is happening, since both state authorities and the Saami people want to sustain reindeer herding. From the Saami point of view, herders need larger herds to maintain their livelihood. The Scandinavian governments, however, have for a decade claimed that pastoralism is unsustainable, and they have wanted to regulate the number of reindeer. Basing their efforts at regulation on the view that Saami culture is tied to usufructuary rights to land and water, the Norwegian state appointed a special commission in 1980, the Saami Rights Commission, to discuss those rights from a strictly legal perspective. The commission also discussed the links between historical practices of land use and environmental issues. In its debate over Saami culture the state faced a tension between its obligation toward indigenous people and environmental protection. After 1987 the Norwegian government made sustainable development the guiding principle of its environmental policy. In theory, supporting human rights would accord with managing nature, but whose standards of sustainability were to be used—those of the politicians or the Saami people? The standards were not necessarily congruent. The government's drive toward more sustainable use entailed legal changes pertaining to property rights and usufruct.

The rights of indigenous peoples received much attention at the United Nations Conference on Environment and Development, held in Rio de Janeiro in 1992. Representatives of Norway, Denmark, and the Nordic Saami Council proposed text relating to the rights of people like the Saami to pursue sustainable development according to their own traditions. Since then the Norwegian government has used Agenda 21 (an international program for sustainable development; see Glossary for more detail) as a framework in its discussions with the Saami. In practice the government passed new legislation in 1996 regulating reindeer herding that curtailed the use of snowmobiles and other motor vehicles in the pastures. Saami areas face new pressure from municipalities that allow the building of recreational homes that are connected by roads and electricity. Grazing areas already under pressure will be further split by these networks, and overgrazing on such small units of land is likely to result. Norway also issued a readjustment program for the interior of Norwegian Lapland (1993–1998) that aimed to reduce the number of reindeer and create other jobs for Saami herders.

VANISHING AGRARIAN LANDSCAPES AND THE CONTRADICTIONS OF TOURISM

In the market-based societies, the ecological and social effects of agricultural modernity left deep imprints on European public opinion. To take but two examples, the destruction of hedgerows in Great Britain and land abandonment in France resonated deeply with their respective national audiences, albeit in different ways. Perhaps half of the hedgerows in the agricultural counties of Britain have fairly recent origins, planted in the era of enclosures from 1750 to 1830 (see chapter 5), yet for many British citizens, the hedgerows and the modest fields they enclose represent the quintessential *natural* landscape. The smiling countryside, nature on a human scale—though these phrases appear to be clichés, they acquired added value during World War II, figuring in national propaganda as something the British soldier was fighting to save. Conservationists appreciated hedgerows as islands of habitat for insects, nesting birds, and burrowing animals. The postwar incentives to produce more food required ripping out hedgerows in order to enlarge the fields. Statistics on hedgerow removal chart, in essence, Britain's postwar agricultural boom: 15 percent of the mid-

twentieth-century stock of hedgerow was removed by 1963, 25 percent was subtracted by 1974, and perhaps 42 percent was gone by 1982. Numbers showing
the loss of fens, heaths, chalk downs, and hay meadows in Britain are even
more impressive, yet the loss of hedgerows most fully represented the demise of
an intimate and familiar countryside. A government-sponsored program to restore hedgerows has been in effect since the early 1970s.

The French equivalent in terms of both dramatic change and its effects on
the national psyche has been, since the 1950s, rural depopulation and land
abandonment. To be sure, the British countryside, too, thinned out in the postwar decades, but depopulation in France drained a much larger, self-consciously
peasant society from the land and into the cities. As high wages in industry exerted their magnetic pull, peasant families were simultaneously pushed by indebtedness, government policies facilitating *remembrement* (land consolidation), and later on a regime of overproduction in which only the largest, most
competitive farmers (who received the lion's share of CAP subsidies) could survive. Between 1950 and 1988, over three million hectares of French soil fell out
of cultivation; by the 1990s, the number of farms had dropped by over half,
down to under one million from two million in the 1950s, and some thirty
thousand farmers were still leaving the land annually. A world that, depending
on one's viewpoint, represented traditional values, a sane and proper relationship with nature, or a cushion against urban alienation was on the brink of extinction. The disappearance of agrarian civilization patently preoccupies the
French government and much of the public.

In West Germany the number of farmers decreased rapidly for the same reasons as in France. West German authorities likewise promoted land consolidation (*Flurbereinigungen*) between the 1950s and the 1970s as one of the cornerstones of agrarian policy. Nearly all arable land was touched, and in many parts
of the country, the landscape changed dramatically. Hedges, clumps of trees, little parcels of fallow land, and areas of extensive exploitation disappeared, all
giving way to monocultures. Land consolidation often went along with road
construction, drainage, and the regulation of rivers. As a result, biodiversity declined, the water balance was altered, and local weather patterns changed. In
some areas soil erosion became an important problem. Since the 1980s ecological science has demonstrated the harmful effects of land consolidation; in many
regions, farmers and authorities are trying to reverse it. Rural areas in East Germany faced the same problems, perhaps in accentuated form. Expropriation and
forced collectivization in the first decades of "real existing socialism" accelerated the concentration and industrialization of agriculture through the Agrarian
Producers' Cooperatives. Many regions of East Germany are still dominated by
factory farms with their endless fields.

What of the ecological consequences for land taken out of cultivation? In France the total surface area of land classified as fallow stabilized in the 1990s, but approximately half of it became lost to agriculture, either from being built over or through transition to dense forest. Some French ecologists and conservation organizations welcome abandoned farmland as potential reservoirs of biodiversity; others decry the return to "wild nature," observing that within twenty years in some regions, farmland gives way to scrub composed of hawthorn, juniper, and broom, vegetation ideal for wild boar but hostile to smaller animals. Paradoxically, biodiversity will have to be created, for example, by introducing rustic species of herbivores capable of maintaining grasslands interspersed with trees—perhaps recalling the Mesolithic mosaic of western Europe (see chapter 1). Land left fallow spells tragedy for many French farmers, who retain strong images of themselves as producers rather than as caretakers of landscape. The term "French desert" now applies specifically to those rural areas marked by abandoned farms and an aging population. Indeed, the steady encroachment of forest—now covering over one-quarter of the French territory—has given rise to the term "green deserts." A century ago, French foresters wrote of actual deserts that appeared to be creeping into the mountainous regions of the country, thanks to deforestation; today, the opposite is occurring, but the common denominator of both eras is human depopulation.

As the farmers left, tourists arrived in the postwar European countryside in general, if not always in those areas worst hit by rural exodus. Paid vacations and the rising tide of car ownership made for a vacation culture that has intensified with each passing decade. Once again, we turn to World War II for critical changes in outlook. In Britain the egalitarianism symbolized by the soldier altered the old debate about access to the countryside. In the "land fit for heroes," writes conservationist David Evans, "all that was best about Britain would be made available to the masses. The landscape would be theirs to enjoy." (Evans 1992, 64) Preservationists such as John Dower also looked to the war's end as an opportunity to pursue the designation of national parks, and ecologists pushed for the creation of nature reserves to save what they feared were the disappearing bits of wild Britain. Hence the National Parks and Access to the Countryside Act of 1949 both set the stage for the creation of national parks (ten would be established in England and Wales during the 1950s) and established the Nature Conservancy, whose brief was to create and manage a national network of nature reserves. Today it manages over 200 nature reserves, although many are small and only under lease to the Nature Conservancy.

Popular clamor to see the countryside did not lead automatically to preservation, however. The north of Scotland is a prime example. By the 1960s economic planners felt that this marginal, depopulated region could be given over to holiday towns, both to service industrial workers and to reinvigorate the Highland economy. An abundance of undeveloped land, opposition from key regional estate owners, and the development of tourism all worked against the creation of national parks in Scotland. Local authorities could cast nature conservation as an imposition by London that ignored local interests. Thus the popular Cairngorms region became a National Nature Reserve in 1954, yet it continued to experience the depredations of deer, sheep, hikers, and skiers. By the 1970s a British family could more easily visit Santa Claus Land at Aviemore than seek isolation in the Highlands. The national parks in England and Wales remained relentlessly subject to local interests themselves. Even where the management of national parks had to conform to strict national statutes, as in France, whose 1960 charter created inner zones of parks off-limits to all forms of economic development, potential profits from tourism threatened preservation. The National Park of the Vanoise, located in the northern Alps of France, was nearly disfigured by a major ski resort, although protest on a national scale convinced President Pompidou to cancel the project in 1969.

The Alps, which arc across seven countries, achieved the undisputed status as an international playground for millions of Europeans and non-Europeans alike: today the Alps are home each year to a mind-boggling 25 percent of global tourist turnover. A new breed, the upper-class mountain climber, "discovered" the Alps in the late nineteenth century for purposes of recreation, but only after World War II did a summer hiking season, followed in the 1960s by a winter ski season, begin to re-create the ecology and economy of formerly agropastoral alpine valleys. Revolving restaurants on mountaintops and skiers dropped by helicopter to the tops of glaciers exemplify how tourism has come into acute conflict with conservation. Forty thousand ski runs; 12,000 lifts; 800,000 trucks and four million automobiles plying the roads annually; disappearing farmers and cows—these and other phenomena linked to tourism have led to more avalanches, floods, and acid rain. Contradictory interests abound: the same "downstream" countries dependent for their water supply on Alpine hydrology, namely Germany, Italy, and France, also have stakes in tourism and in the trucking of cargo. It is the geographical fate of the Alps to be centrally located in western Europe, and the integration of European economies has made the world's busiest mountain roads only busier.

A spectrum of ecologists, public officials, and common citizens came to recognize by the 1970s that mass tourism might be detrimental to fragile habitats, especially alpine regions. By then, northern European countries had state-level conservation bodies, some of them, such as Britain's Nature Conservancy and the Netherlands' Department of Nature Conservation, dating from the

1940s. France led in creating an actual Ministry of the Environment (1971), although only gradually did it acquire sufficient authority and funding to begin to deliver on the promise of environmental protection. Meanwhile, Europe became implicated in a web of international activity, beginning with the United Nations–sponsored Conference on the Human Environment, held in Stockholm in 1972. The conference prompted leaders of the European Economic Community to launch their own Action Programme on the Environment, and the EEC soon began to issue directive after directive pertaining to environmental policy. The 1980s saw the concept of sustainable development slowly gain international standing, first through the World Conservation Strategy, spearheaded jointly by the United Nations and the International Union for the Conservation of Nature and Natural Resources, later and more decisively through the report of the United Nations' Brundtland Commission, *Our Common Future* (1987). The European Union's Treaty of Amsterdam (1997) sought to bring principles of sustainable development to the center of all aspects of policymaking within the EU (see case study, "From Nature Conservation to Sustainable Development: The Scandinavian Experience" on pages 188–203).

ENVIRONMENTALISM: FROM GRASSROOTS ACTIVISM TO THE FORMATION OF GREEN PARTIES

By the 1970s, all of these initiatives were being pushed from below, by individuals, organizations, and media attention, all of which helped catapult the modern environmental movement. In contrast to the older, often well-heeled conservation and preservation movements (see chapter 5), a new militancy led to different forms of organization and practice among environmentalists. A host of cultural and political changes nurtured the movement: growing qualms about technological modernity and consumer society; a more outspoken stance on the part of scientists, who now more than ever voiced the risks associated with scientific discoveries and applications; and the sense of democratic renewal that burst forth from the student movement of May 1968, based in Paris. Even though an environmental agenda had been noticeably absent from the clamorous demands of the students, their insistence on political decentralization and grassroots power came to inform environmentalism in the 1970s. Something else was present from 1968 onward—the notion, best expressed in French, of a *choix de société*, the idea that all of the determinants of modern society, from consumerism to managed democracy, were now up for grabs. The future was to be chosen rather than shaped by abstract forces.

Detergents sprayed at Porthleven, England, to disperse oil from the Torrey Canyon *spill, March 26, 1967. The detergents proved fatal to many of the birds that survived the oil itself. (Bettmann/Corbis)*

Finally, the media contributed to a general public consciousness of rare, wondrous, or threatened habitats. By the late 1950s the BBC was producing no fewer than five nature shows for television. Underwater explorer Jacques-Yves Cousteau produced his first major film in 1956; over several decades he succeeded in revealing a new subset of wilderness, the ocean depths, to millions of armchair explorers. Cousteau's films also documented the growing load of garbage and other forms of pollution in the world's oceans, in addition to the ravages of unchecked fishing. The media were beginning to report on dramatic environmental "events," too. For Europeans, the seminal environmental disaster of the 1960s was the grounding of the oil tanker *Torrey Canyon* on a reef off of the Isles of Scilly (in the United Kingdom) in March 1967. Over twelve days it released its cargo of 860,000 barrels of crude oil, soiling the coasts of Cornwall, Brittany, and Guernsey. The televised images of oil-covered seabirds, some 25,000 of whom died due to the oil slick or to the toxic detergents used to clean

Greenpeace activists staged a protest next to the 26-year-old single-hull petroleum tanker Byzantio *on December 5, 2002, in Hoek van Holland, the Netherlands. The Maltese* Byzantio *carried 70,000 tons of petroleum and was one of 2,200 blacklisted dangerous ships. (Getty Images)*

beaches, brought to light the hazards of the substance, crude oil, that underpinned much of the European economy.

New international organizations committed to direct action also commanded public attention. Greenpeace and Friends of the Earth, both founded in 1969, spawned branches in most European countries within a few years. These two groups set about promoting recycling and cleaner forms of transportation while protesting whaling and nuclear testing with characteristic panache and (notably in the case of Greenpeace) quasi-legal tactics. Both remain well supported in much of northern Europe: the Dutch branch of Greenpeace, the largest in the world, had attracted 700,000 members by 2002. In a broader context, nearly one-quarter of the Dutch population belongs to some kind of conservation or environmental organization—one of the highest rates of public participation in environmental causes in the world.

Antinuclear activism best characterized environmental militancy in the 1970s. The fight against civilian nuclear energy programs brought together many activists because of the industry's wide social as well as environmental implications. Nuclear energy required technological gigantism and a technically trained elite that could skirt democratic control. The danger of accidents, the threat of coolants raising the water temperature in rivers or altering local climatic conditions if released as steam, and the thorny issues surrounding radioactive waste were problems that flew in the face of environmentalist values centering on small-scale democracy and appropriate technology.

Comparing the French and German nuclear programs highlights the importance of very specific national circumstances in determining the fate of nuclear power in two equally oil-poor countries. The case of France, a country "going nuclear" rapidly after 1974 (thanks to the Plan Messmer), shows that a high-stakes technological project underwritten by the state outweighed the determination of activists. France's *force de frappe*, the nuclear arsenal born in the 1960s, actually enjoyed increasing public support in the following decade, as did civilian uses of nuclear energy; within Europe, France was the "lone star in the statistical firmament" in terms of pronuclear public opinion. (Bess 2003, 99) The antinuclear movement in France did attract scientists, who published their appeals in *Le Monde*, as well as jurists and dissidents from the nuclear establishment, but it also faced daunting opponents and committed tactical errors. The first demonstrations of the French antinuclear movement were peaceful, for example, its opposition to a nuclear reactor at Fessenheim (Alsace) in 1971. But in subsequent violent clashes with police at Super-Phénix in 1977 (the "jewel in the crown," a fast-breeder reactor near Lyon) and again in Brittany in 1980, the movement adopted increasingly confrontational tactics and became easily typecast as extremist by the media. The movement suffered its supreme setback in the early 1980s, when the newly elected Socialist government under President François Mitterrand opted to pursue France's nuclear program.

From the early days of atomic power, the West German government supported physicists advocating a national nuclear industry. Chancellor Konrad Adenauer likely wanted to assure West Germany's access to nuclear weapons, but a West German nuclear force was never realized. Private energy suppliers even kept their distance from schemes to develop civilian nuclear energy. By the late 1950s they remained unconvinced that atomic energy was good business. In contrast to the industry's skepticism, the political class and the media expressed atomic optimism. At the famous "atomic party congress" of the Social Democrat Party in 1956, several speakers outlined their vision of a future world that would be supplied by free energy. Many contemporaries were convinced that nuclear power would cost little. Cheap energy would enable the reduction of working hours and the industrialization of the Third World, while nuclear byproducts could be used in treating serious illnesses such as cancer.

Fueled by sanguine expectations, the West German state promoted nuclear energy and financed the necessary research programs. Political pressure and subsidies handily persuaded the energy industry to get into the nuclear business during the 1960s. The first commercial nuclear power plant was inaugurated in 1968, with several other atomic power stations following in subsequent years. More than twenty plants were under construction or in the planning stages by 1972, but the expansion of nuclear power did not last for long. The last order for a plant was given in 1982, and the last reactor went online in 1989.

The quick fall of West German nuclear power resulted from two circumstances. On the one hand, early studies predicting a dramatic increase in energy demand—and its purported cheapness—had to be corrected. On the other hand, the protests of a powerful antinuclear movement were increasing the political costs of nuclear power by the middle of the 1970s. The West German antinuclear movement united very different groups, from conservative farmers to middle-class women and radical students. Nuclear power became a symbol of outsized technology and arrogant administrations that ignored the interests of local inhabitants. Writers like Robert Jungk suspected politicians of using atomic energy as a pretext for establishing a police state. Indeed, West German officials reacted vigorously to antinuclear protests, and in the second half of the 1970s several demonstrations ended up in violent clashes.

Beyond demonstrating, antinuclear activists used the opportunities provided by German administrative law. By commissioning critical scientific expertise and filing lawsuits against permits, they managed to draw out the planning processes and to force up costs. The antinuclear movement tallied its successes in part because some of the scientific and judicial elite supported its claims, as had occurred in France, but in Germany the mass media also turned against nuclear energy when antinuclear activists opted exclusively for nonviolent protest at the end of the 1970s. Several developments of the 1980s sealed the fate of the German nuclear industry: the new ecological party, the Greens, represented the struggle against nuclear power in the national Parliament, and in 1986, when Chernobyl's atomic power plant exploded and a radioactive cloud drifted across the German sky, massive antinuclear demonstrations broke all records. Politics compelled the conservative Chancellor Helmut Kohl, who advocated nuclear energy, to establish a Ministry for Environment and Nuclear Reactor Safety. Antinuclear protest concentrated on the transportation of atomic waste in the 1990s. When the Green Party and the Social Democrats formed a ruling coalition in 1998, the legacy of the German antinuclear movement became a principle of government policy. In 2000, the government and the energy industry signed an agreement stipulating the closure of the nuclear energy program.

By the 1980s European environmentalism was moving away from face-to-face contestation, epitomized by the antinuclear movement, and toward a partial integration with mainstream culture. One might perceive this trend as a shift from "outsider" to "insider" status: many organizations turned to more conventional activities, while individual environmentalists began to lend their expertise to commercial and government projects. Alternatives to the main systems of production—organic agriculture, wind energy, and recycling, to name just a few—that were once marginal became commercial enterprises. Policymaking at many levels now routinely responded to environmental concerns. Consumerism began to acquire a green veneer, as ecolabels on products, catalytic converters in cars, green tourism, and other forms of ecoconsumerism took off. The economic indicators of these decades reveal only the veneer that coated both government policy and commerce. These were compromises integral to what Michael Bess terms the "light-green" society of postwar industrial democracies, "an increasingly pervasive overlay of ecological ideas and environmental constraints upon the growth-driven, consumer-oriented system inherited from mid-century." (Bess 2003, 241) Card-carrying European greens themselves remained divided over whether the consumer economy in their midst was best dealt with by promoting reform or through radical alternatives.

Meanwhile, the word "green" was acquiring a capital "G": the environmental movement began to come of political age as Green parties sprouted across the European map. French environmentalists had convinced well-known agronomist René Dumont to run for president in 1974; in hindsight, a presidential candidate running with a green agenda was far more significant as a political turning point than indicated by the mere 1.32 percent of the vote that Dumont won in the first round of the election. Novel political stances that valued ecological health over economic growth led to the creation of new parties. New Zealanders established the world's first Green Party in 1972. From 1973 to 1988, citizens in all of the countries covered in this book founded and gained official recognition of national Green parties. The strong Belgian Green Party achieved representation in the Belgian national Parliament as early as 1981. French Greens made their electoral breakthrough in 1989, entering a number of municipal governments as well as the European Parliament. Disadvantaged by an electoral system that is relatively unfavorable to small parties in national elections, French Greens had to wait until 1997 to enter the National Assembly. (On Germany, see the case study "The German Green Party" on pages 175–188.)

Despite their reputations for being in the forefront of environmental policy, the Scandinavian countries, like France, lagged behind England and Germany in spawning formal Green politics. Sweden's "Environmental Party," founded in

1981, was the offspring of a referendum on nuclear energy and a groundswell of criticism of ordinary parties for not taking environmental issues seriously. In 1982 the Environmental Party failed to garner the 4 percent of the vote necessary to enter Parliament, but some candidates did get elected to several municipal boards around the country. In order to accent their resemblance to other Green parties in Europe, the Environmental Party added "Green" to its name in 1985. Honing their focus on the interrelationships among social, economic, and environmental issues, Swedish Greens based their ideology on four principles of solidarity: solidarity with animals and ecosystems, solidarity with future generations, solidarity with all people in the world, and solidarity with people in Sweden. The party scored better in the second half of the 1990s as anti-EU voters lent their support. The Swedish Green Party has never entered government, but it has cooperated with the Social Democrats and the Socialist Left in Parliament.

Norway was a latecomer to the Green party phenomenon. In local elections in 1987, a group of academics presented themselves as "the Greens" and won seats in some small towns around Oslo. Encouraged by this initial success, they formed a nationwide Green Party in 1988. However, the Greens attracted few enthusiastic voters and came away with a tiny 0.5 percent of the vote in their first general elections. Whereas the Swedish anti-EU voters supported their national Green Party, this did not happen in Norway. Anti-EU voters sided with the Socialist Left and the agrarian Center Party. These two parties were already waving the banners of green values, local democracy, rural lifestyles, and national sovereignty; elements of Green party platforms elsewhere in Europe had other political homes in Norway. To this day the Norwegian Green Party has not taken a seat in Parliament.

Overall, European Green parties have proven to be a stable element in politics. Electoral support at the national level has ranged from 1.5 percent to 7.3 percent, results that often please the many Greens who do not wish to belong to a large party. The late 1980s brought more voters into the Greens' ranks. Diverse factors were at work, including the experience of the radioactive cloud that blew north and west from the Chernobyl meltdown of 1986, increased media coverage of the greenhouse effect and the destruction of the ozone layer, and, not least, a respite in the economic downturn that had plagued Europe since the early 1970s. Europe reflects a historical, and global, trend according to which environmental ideas often receive more of a hearing in conditions of relative economic prosperity. The strongest results for Green parties of the last decade have occurred in Austria, Belgium, Finland, Germany, Luxembourg, and Switzerland. For many European voters, though, Greens have yet to prove that theirs is not a single-issue party and that they are capable of taking on weighty

German Foreign Affairs Minister Joschka Fischer speaks during a meeting in Rome on February 20, 2004, to launch the European Green Party, the first Europe-wide political party. (Tony Gentile/Reuters/Corbis)

political responsibilities. Greens have entered government in Finland (1995), France (1997), Germany (1998), and Belgium (1999), at times receiving crash courses in the exercise of power.

FORMING THE FUTURE: SUSTAINABLE AGRICULTURE, HYBRIDITY, AND THE ELUSIVE EUROPEAN WILDERNESS

As the original core of the industrialized world and now the globe's largest economic bloc (our measurement being the Gross Domestic Product of the European Union), Europe contributes mightily to problems that afflict the entire planet in different ways and degrees, from global warming (originally theorized in the late nineteenth century by the Swedish chemist Svante August Arrhenius) to ozone thinning and loss of biodiversity. Northern Europe has its own "hot spots" that seem to defy solution, such as the North Sea, forced to absorb 70 million tons of pollution annually, in the form of dredge spoils, agricultural runoff, untreated sewage, radioactive cooling water, oil, and acid rain. Europeans have been well instructed in nature's reactions to such an onslaught. Let us conclude with an overview of ideas and directions in the conservation of European nature as a whole.

Forestry, for some a venerable pillar of the conservation movement, has shaky environmental credentials in the eyes of others. The planting of trees, or entire forests, seems to herald ecological responsibility, as forests soak up carbon dioxide, regulate the flow of water, help prevent erosion, and provide a renewable resource. Seen through another lens, planted forests sometimes replace native biomes and appeal little to Europeans' aesthetic sensibilities. The British Forestry Commission's gradual afforestation of 400,000 hectares of blanket bog in northernmost Scotland, the "largest mire in the world and the most extensive piece of single habitat in Britain," grew highly politicized even as 60,000 hectares had become forest by the end of the 1980s. (Evans 1992, 202) The largely deforested Republic of Ireland has seen aggressive afforestation programs since 1950; the government's goal of a half-million additional hectares to be planted over the first third of the twenty-first century—bringing the total area of Irish forest to about 10 percent of the national territory—would yield a purely industrial forest adapted to the needs of the timber industry. Such forests have, however, already proven their fragility in the face of disease and high winds. The extraordinary force of Hurricanes Lothar and Martin, which swept across northern France and Aquitaine in December 1999, revealed images of trees downed like matchsticks. Within nearly a million hectares of damaged forest, it was coniferous high forest, managed for density, uniformity, and height, that proved most vulnerable. France's National Forest Office has taken a markedly soft, relatively noninterventionist approach to replanting.

What, then, is the "right" nature to live in and preserve? Such a question takes us back to the ancientness of human manipulation of ecosystems in Europe. Today the largest "lake" in northern Europe is a human artifact, Keilder Reservoir in Britain's North Tyne Valley, located within the largest planted forest in the United Kingdom. Fundamentally, though, the indigenous European notion of "right" nature tends toward the rural and agricultural. This sweeping generalization takes into account not only the visions of Greens themselves, recognizing the handiwork of several hundred human generations, but also of a much larger group of Europeans who repeatedly equate their preferred "nature" with the countryside when asked about such ideas in opinion polls and surveys. Much of our region, from southern Scandinavia to the Alps, remains under cultivation or else is highly marked by agriculture, yet as we have seen, the post–World War II model of farming devoured biocides, fertilizers, and water while producing surpluses at a predictable environmental cost.

That model is, officially, no longer in place. From the early 1990s the European Union reduced price supports and devised set-aside schemes by which large producers are paid not to cultivate given portions of their land. A major reform of the CAP took place in 2003. Subsidies once linked to production levels will gave way, from 2005 on, to single payments tied to farmers' adherence to standards regulating food safety, environmental protection, and the welfare of farm animals. Encouraging traditional crops and methods, soil protection, and rural tourism are all part of the EU's vision to steer European agriculture away from five decades of intensification. The CAP budget will be gradually reduced after 2006, despite the entrance of fifteen new states from eastern and southern Europe—home to less advanced but also less environmentally destructive agriculture—in 2004. Tellingly, the reform confirms the deep European association between nature and agriculture, for in the future, the EU is likely to handle matters pertaining to agriculture, environment, and sustainability under a single policy rubric titled, "preservation and management of natural resources."

On paper, this far-reaching shift in policy would seem to please many European Greens, perhaps even the iconoclastic José Bové of France's Confédération Paysanne (Peasant Confederation). Concrete results are likely to be as varied as the diverse agricultural regions that lie within the European Union. Looking beyond field and pasture, environmentalists identify other ecosystems that are vital to preserve or restore. Europe cannot, despite its many national parks, claim to have wilderness areas of the breadth and relative wildness that one finds in

Australia, Canada, or the United States. Rather, Europe is the home of hybrids—places that partake of both tame and wild qualities, ranging along a spectrum from the relatively untouched to the relatively artificial. Though one might argue that hybridity now characterizes planet Earth for all practical purposes, it is clearly within this framework that most discussion about conservation and preservation in Europe takes place. To quote conservationist David Evans again, "Much of what we seek to preserve in the way of habitat and landscape has resulted from our own activities. The woodlands and the coppices, the pastures and the hedgerows, the parks and orchards and many of the lakes—all have been made by the hand of man. They are tenanted by animals and plants that can adjust to change, given half a chance." (Evans 1992, 257)

To provide the framework does not yield the answers, however. Two examples of controversy, very northern European in their flavor, will suffice. First, it is not clear whether some of the worst instances of industrial blight should undergo ecological restoration. Does artifice itself, though it once blatantly violated the ecology of a place, deserve preservation for the sake of historical memory? Some industrial relics, such as Britain's largest slag heap at Cutacre Clough, Lancashire, have achieved notoriety and a firm place in local culture; even conservationists have argued on behalf of the plants and animals that have recolonized such sites and would be disturbed by "restoration." As a counterexample, beginning in 1990 a new and extensive forest was planted on industrially derelict land in the Midlands, an undertaking which received much popular support and seems to exemplify multipurpose forestry and habitat creation. This leads to our second controversy: how legitimate is the creation of new habitat? In the early 1990s the World Wildlife Fund announced the ambitious goal of doubling the "real nature" in the Netherlands from 200,000 to 400,000 hectares within twenty years. In a country like the Netherlands, "real nature" can only mean a highly managed landscape requiring the most purposive stewardship. For some conservationists, this signals a laudable undertaking in one of the world's most densely populated countries; for others, new habitat smacks of "gene banks," necessary yet lacking aesthetic appeal and reflecting the failure of conservation in its older guise. Catastrophic declines in bird, butterfly, and native plant populations in Britain over the last thirty to forty years, catalogued in March 2004 in the journal *Science*, heighten the urgency attached to both preserving and creating habitat.

There may yet be room for the preservation of nearly wild nature in northern Europe. F.W.M. Vera, author of *Grazing Ecology and Forest History*, would allocate far more space for the proliferation of wild flora and fauna. His vision calls for reintroducing large herbivores into forests and abandoned farmland, even "rewilding" those varieties of domestic cattle and horses that have not been overbred. The large herbivores that lived during the Paleolithic helped create the particular mix of grassland and forest that is lowland Europe's most characteristic biome (see chapters 1 and 5). These animals, reintroduced by humans, can re-create and maintain those rich and resilient ecosystems. Championing room for wild nature in cultured Europe, Vera asserts, "It is only by knowing the wilderness that we can understand our cultural landscape." (Vera 2000, 384) Yet human design will be profoundly enmeshed in the "new wilderness," should it ever be created.

If politics will lead the way in fostering sustainability in Europe, the inevitable role of the European Union raises questions for northern Europe and its neighboring regions. Will political and economic integration help foster a vision of the unity-within-diversity of the European northern temperate zone from an ecological perspective? As more initiatives pass to the EU, will nation-states and, more importantly, their citizens inform and abide by the evolving panoply of environmental rules and regulations? If rules, regulations, management, and monitoring fall short of their makers' goals, as they are likely to do, ordinary Europeans will be called upon to launch the cultural shifts necessary to repaint their societies' shade from light green to a deeper verdure.

Notes

1. The phrase comes from the thesis of Garrett Hardin, expounded in 1968, by which common or global resources will become degraded as individuals act according to their own interests. See the Glossary for more detail.

2. The Saami people have inhabited northern Fennoscandia since approximately 1500 BCE. They live as far south as Røros in Norway but mainly inhabit the northern parts of the Scandinavian peninsula as far east as the Kola peninsula in Russia. The Saami Parliament replaced the old term "Lapps" with "Saami" in 1973. According to Saami law in Norway, a Saami is someone who either considers him- or herself to be a Saami, speaks Saami as a first language, or has at least one parent or grandparent who learned Saami as his or her first language.

7

CASE STUDIES

CASE STUDIES: FUEL RESOURCES AND WASTELANDS IN THE NETHERLANDS AROUND 1800

Introduction

Peat ["*turf*" in Dutch] is the fuel of our fatherland; Don't use too much
  of it.
Peat is part of the soil we live on.
T t (Turf)
Peat is a warming, clean, durable and in our region healthy fuel.
It is certain, that we steadily burn a part of our soil
and because of our growing wealth, more peat is used today than in
  more restricted times.
It is also certain that neither low, nor high peat will grow in volume
  again,
or at least, this growth will not be significant.
So, we know that our peat reserves will diminish every year.
Should we not try to compensate this by planting trees?
Most of our dunes and many other places not fit for agriculture or cattle breeding
are very fit for the planting of trees.
If we did not drain off the dredged peat bogs, wouldn't that be a dangerous and harmful shortcoming? Nobody can live in lakes and
  marshes, and the power of the state rests on the size of its population.

Vaderlandsch A-B Boek (Patriotic A B Book) 1781

In the late eighteenth century, there was much concern in the Dutch Republic about the disadvantages of the use of peat as a fuel. Peat was used in many industries, such as brick making, sugar refining, distilling, brewing, and more.

According to at least one historian, the Dutch Golden Age was born from peat. Peat was also the most important fuel for cooking and heating in many parts of the country where not enough wood was left after the serious deforestation in the Middle Ages. This deforestation was not without danger, as sand drifts could, as elsewhere in Europe, destroy arable land and farms.

In England in the early modern period, after serious deforestation, wood and charcoal were replaced by coal, which caused much air pollution. Peat, however, was a rather clean fuel. Although the heat of this fossil fuel was not enough for the melting of metal ores, it was very useful. In the nineteenth century it was even used to drive steam engines.

The problems in using peat were twofold: the country would eventually run out of peat, and the dredged peat bogs in the west of the country would turn into lakes. The reason behind this development was that the most important peat areas barely rose above, or even lay beneath, water levels. As soon as they were dredged, lakes formed or deepened. Winds would later cause extensive land erosion on the borders of the lakes, which grew considerably. Farmland disappeared, and in some case entire villages vanished. This problem of land erosion and resource exhaustion has a decidedly modern ring to it. The destruction of the land by mining—for instance, the digging of lignite in the former German Democratic Republic—is one of the environmental problems Europe has to deal with. The formation of large lakes due to erosion can be compared with the damage caused by another, more modern way of obtaining energy: I refer here to the ever-growing reservoir lakes, for example, in the former Soviet Union.

The above text reminds us also of another modern problem: the probability of running out of fossil fuels. This problem became familiar to us after the publication of *Limits to Growth* in 1972, the first report to the Club of Rome, but the late-eighteenth-century text has even more modern connotations. Substituting fossil fuels with sustainable energy sources such as solar power, wind power, hydro power, or the burning of biomass is nowadays often promoted in order to obviate a possible energy shortage in the future. The Dutch environmental scientist Hans Opschoor suggested ten years ago that in a sustainable economy, the "use of fossil fuel would be permitted if equivalent biomass or solar energy devices were put aside for future generations." A most remarkable part of the introductory text is that two hundred years ago, in the Dutch Republic, a solution—the planting of trees on fields not fit for agriculture—was already envisioned as a way to solve the energy crisis that would develop once the peat resources were exhausted.

At the same time the text quoted above offers a solution to the problem of the formation of lakes and land erosion: they could be drained, by using wind power, another sustainable energy source, and turned into polders pro-tected by dikes.

Even more remarkable was that both solutions were really accomplished over the next century: many polders were created, and large tracts of otherwise useless land were planted with trees. Within a century or so, the Dutch man-aged to solve the problem of lake formation in the west of the country and to make better use of derelict heath and sand dunes, along the coast as well as in-land, by afforestation. This was so successful that today, the last remaining lakes and the last bare dunes and heaths are protected by law, for they are used for recreation, for fishing, as a source of water for the cities, and for natural parks. The forest plantations of the nineteenth century are not necessary for fuel anymore and are nowadays very popular with tourists. The famous Kröller-Möller Museum, containing many paintings by Vincent Van Gogh, lies in the middle of extensive tracts of forest where two centuries ago there was nothing but heath and sand drifts.

Viewed from an environmental perspective, this seems to be a success story. But before analyzing the matter further, one should first take a glance at the last clause of our text: "the power of the state rests on the size of its popula-tion." The writer was not referring primarily to an environmental problem, but to a political one. By the late eighteenth century the economic and political power of the Dutch Republic had dwindled, population growth had stagnated, and France and England had become far more powerful. A war with England in the years 1780–1784, in the aftermath of the American War of Independence, proved disastrous. Widespread concern about political weakness catalyzed many proposals to restore the power of the Dutch state. The late-eighteenth-century text in the *Patriotic A B Book* has to be seen in the light of physiocratic ideas and a patriotic ideology purporting that effective land use was the best way to build a strong nation. Nevertheless, ecological problems existed, caused by a self-inflicted expansion of wastelands.

First, the extent of the problem of lake formation and the ways in which it was tackled will be analyzed. Second, the problem of fuel resources should be considered. Third, this case study will focus on the process of afforesting the wastelands. In the early modern period European governments became inter-ested in solutions to environmental problems and land use. What matters most, therefore, is the question of whether we are dealing with politically inspired so-lutions: were governments able to solve the problems, or was the success story just a matter of coincidence?

To gain a more in-depth understanding of the questions posed above, two examples will be analyzed. The small province of Utrecht in the center of the Republic, which was extremely hard hit by land erosion, will be used to illustrate landscape erosion, lake formation, and dwindling fuel resources. The specific role of the state will then be examined through the attempts of the Dutch government in the nineteenth century to promote the planting of trees in one of the most severely deforested regions in the country. These were the state-domains in the western part of the Veluwe, just east of Utrecht and about seventy-five kilometers away from the capital, Amsterdam. These domains were very hard hit by the extension of sand drifts.

In Dutch economic historiography, these two subjects—the creation of polders in former peat regions and afforestation—are normally treated separately. However, from the point of view of environmental history they were strongly related and both had to do with the physiocratic view of land use. Both projects flowed from the presupposition that lakes and "wasteland" were useless. They were both considered dangerous, as well.

Low-Peat Digging, Lake Formation, and Drainage in the Province of Utrecht

INTRODUCTION.   In the Netherlands, one can find two different forms of peat: high peat and low peat. The difference lies not in their geological formation or substance but rather their location. Low peat lies close to or beneath sea level and was mostly dredged from under the waterline. After being dredged, the peat was dried on parts of land spared out between the dredged pools. High peat, on the other hand, was dug from peat formations situated on higher ground. The legacy of cutting high peat was quite different from that of dredging low peat. When high peat was dug out, and the fertile top layer of soil had not been preserved, the soil that came to the surface was generally unproductive for agriculture. It had to be fertilized with dung from the cities to regain its fertility. Utrecht was a low-peat area and as such was simply lost to agriculture because of the formation of lakes.

Both the formation of lakes due to the dredging of peat and the erosion of lakeshores had serious consequences. In the first place, arable land and pasture were lost, and what remained could only be used for fishing. From the authorities' point of view, this meant that although fishing rights were leased, no further land tax could be levied. Second, buildings were destroyed, and in the neighboring province of Holland, parts of villages had to be abandoned as well because of the ever-growing lakes. Finally, the areas became depopulated. Once the peat had been dredged, no work remained. In some places this last problem was more serious than in others. Northern regions shipped large quantities of peat to the large city of Amsterdam, home to more than 200,000 inhabitants in

TABLE 1
Tax Yield in Guilders on Peat Digging in the Province of Utrecht (Dutch Republic), 1582–1801

| Period | Region | | | | | | | Total |
	1	2	3	4	5	6	7	
1582–1601	796	235	0	218	163	209	0	1,621
1602–1621	1,198	2,196	317	1,355	186	188	0	5,440
1622–1641	639	2,322	1,659	1,383	306	545	0	6,853
1642–1661	127	3,355	1,653	1,360	805	959	0	8,260
1662–1681	76	1,943	778	888	534	986	0	5,205
1682–1701	12	1,177	626	1,016	541	1,172	66	4,610
1702–1712	n.a.							
1712–1731	0	683	308	1,494	645	2,127	113	5,371
1732–1748	0	1,120	288	2,306	651	1,731	308	6,404
1749–1798	n.a.							
1799–1801	0	n.a.	377	50	1,508	2,328	2,127	> 6,391

Source: S. W. Verstegen, *Gewestelijke Financiën* [Provincial Finances], *Utrecht 1579–1798* (The Hague: Institute for Dutch History, forthcoming).

the eighteenth century. Not surprisingly, these peat regions became more quickly exhausted than in other areas where peat was destined for smaller towns, such as Utrecht with its roughly 30,000 inhabitants. Nevertheless, as time went by some exhausted peat regions in the province of Utrecht also lost population and witnessed a rise in poverty and poverty-related crimes.

RUNNING OUT OF RESOURCES? At a rather early date, authorities in the province of Utrecht tried to slow down the process of peat dredging by taxing every parcel of peat moor that had been dredged. In 1582 a special tax on soil removal came into effect. Despite this tax, peat production was so profitable that it did not obstruct dredging in any observable way. Thanks to tax records, it is possible to analyze to what extent peat resources were running out over the course of two centuries (see Table 1). The tax yields give only a very rough indication of the process of peat digging, because many complexities in the collection of the tax remain invisible in the sources. Nevertheless, as far as we know, the tax rate remained constant, allowing for accurate comparison across many decades.

Detailed information is available for seven regions in the province over the years 1582–1701, 1712–1748, and 1799–1801.

Table 1 clearly shows how in one region after another, peat digging started, reached its zenith, and slowed down. In region 1, peat digging reached its zenith

TABLE 2
Tax Yields in Guilders from the Export of Peat in the Province of Utrecht (Dutch Republic), 1700–1774

Period	Export Peat	Indices (1700–1708 = 1.00)
1700–1708	2,105	1.00
1709–1716	1,404	0.67
1717–1724	2,098	1.00
1725–1732	1,174	0.56
1733–1740	1,733	0.82
1740–1748	5,044	2.40
1751–1758	11,921	5.66
1759–1766	14,214	6.75
1767–1774	14,516	6.90

Source: S. W. Verstegen, *Gewestelijke Financiën* [Provincial Finances], *Utrecht 1579–1798* (The Hague: Institute for Dutch History, forthcoming).

in the early seventeenth century and became a marginal affair within a few decades. Peat digging in regions 2 and 3 reached its zenith when peat digging had almost stopped in region 1. In region 4, peat digging developed quickly and remained important until the middle of the eighteenth century. Peat digging in regions 5, 6, and 7 was not significant in the early seventeenth century, but these regions led the effort at the end of the eighteenth century. Though some regions clearly fell out of use, until the mid-eighteenth century there was no serious threat of resource exhaustion, for tax yields between 1715 and 1750 were rising. A half-century later, the total tax yield still does not show a downward turn. The diminishing tax revenues after the middle of the seventeenth century should not be seen as a sign of resource exhaustion; rather, it mirrors the economic downturn in the province which set in after the French occupation in 1672.

Another tax that did not stop the process of peat digging was an export tariff introduced in 1679. The tariff on the export of peat remained constant between 1699 and 1775. As can be seen in Table 2, exports rose in the eighteenth century. In these sources, too, no exhaustion of resources can be diagnosed.

Earlier research has shown that in the nineteenth century peat production in the province of Utrecht continued until midcentury, diminishing suddenly thereafter. The abolition of the tax on coal, not the exhaustion of peat, caused this decline. With the modernization of Dutch infrastructure after 1850, the entire picture of energy resources in the Netherlands changed as it became possible for large quantities of coal to be imported from Germany. By that time, coalfields had also been located in the southern province of Limburg. An absolute lack of energy resources never became a problem.

FINANCING DRAINAGE.   Solving the problem of lake formation by turning lakes into polders was already practiced in the early seventeenth century. However, in the province of Utrecht, drainage was not undertaken until centuries later. Crucially, the provincial government forbade the dredging of peat unless a plan was set up to create a polder in the demolished area. Six large polders in the province of Utrecht were created, using steam engines, between 1789 and 1872, comprising about forty-eight square kilometers. In the nineteenth century the Dutch government advocated the creation of polders but was less prone, due to financial straits, to invest money in such projects. In Utrecht, however, an alternative way of financing polders comprised a history all its own.

As mentioned above, the formation of large peat lakes caused revenue from the land tax to decline. To compensate for the losses, the province of Utrecht had established a fund in the late sixteenth century. From 1592 onward, investors who wanted to produce peat were obliged to pay a certain amount of money as a deposit on a unit of land. This innovative method of raising revenue amounted to fifty guilders per *morgen* (a small hectare). The tariff was doubled in 1694, doubled again in 1696, and was raised to 300 guilders in 1736. In that year investors could no longer reserve land as a security. This money was then lent out to the government, and the interest was used to cover the land tax. Authorities justified the rate increase of 1694 by the higher rate of the land tax, whereas in 1736 officials raised the rate because interest rates on government loans had fallen. It should be noted that the higher land tax was needed to finance the wars of the Dutch Republic against Louis XIV. In a rather ingenious way, then, the province solved a financial problem that had been caused by environmental degradation. The amount of money involved was about a half-million guilders in 1759 and at least 590,000 guilders in 1811. In the nineteenth century this institutional arrangement to finance the making of polders underwent a structural change. In 1836 the money had to be used to finance the drainage prior to creating polders. The state could thereby stimulate the process of landscape restoration while the money came from elsewhere.

Through the creation of new polders, much damage caused by peat digging was rectified. However, drainage ceased around 1900, long before all the lakes were gone. Part of the money in the funds was not even used for drainage, and some 1.7 million guilders of "working capital" remained after World War II.

The creation of new polders stopped in part because of the high costs involved in keeping them dry. Moreover, popular appreciation of the remaining lakes grew for environmental and recreational reasons. Nature conservationists succeeded in halting the process of drainage. In 1942 the lakes came under legal protection.

Deforestation and Afforestation in the Veluwe

SAND DRIFTS.   Two centuries ago, only 4 percent of the Netherlands was covered by forest. The most important woodlands consisted of coppice. Most forests in the Dutch Republic had been felled or degraded in the Middle Ages, and large tracts of former woodland had changed into heath, shrubs, and sand dunes. The problem of drifting sands grew serious in the Veluwe, the area scrutinized in this case study. As early as Carolingian times, deforestation began at the time of the discovery of iron ore deposits, which were rich enough to sustain a weapons industry. The charcoal was supplied, of course, by local forests. Burning the forests to create arable land, overgrazing, and intensive cutting of sod created sand drifts that became a nuisance in the sixteenth century; the local government appointed a functionary to handle the joint problem of sand drifts and burning heath in 1532. The first ordinance concerning heath burning and sand drifts dates from 1566. It prohibited the grazing of sheep on the sand-drifted areas. In 1650 the problem remained on the agenda, but local villages, lacking the means or the will, refused to cooperate. In 1691 detailed regulations stipulated how the sand drifts should be curbed. Sod was to be spread out on the sands, and grasses and coppice were to be planted. A century later, there were continuing complaints about arable land lost under heaps of sand, but they spurred few people to act. Extreme droughts exacerbated the problem at the turn of the nineteenth century.

The political problem was that the peasants in this sandy region had a truly marginal importance for the economy and public finance. Moors were in practice exempt from the land tax, and, in contrast to the lakes, sand drifts were located in one of the most backward areas of the Dutch Republic. The most important economic activity here was sheepherding, but the herds had diminished since the sixteenth century.

In the first half of the nineteenth century, sand drifts covered 10,659 hectares, approximately 10 percent of the Veluwe region. The most extensive were more than 2,000 hectares in area by 1852. Using seventeenth-century techniques, local and provincial authorities accomplished modest objectives. Fast-growing pines, however, had replaced coppice as the favored tree species to keep the dunes in place. In 1900 sand drifts still covered 9,927 hectares. Only in the twentieth century were the drifts really pushed back; today about 900 hectares of them remain.

RECLAIMING WASTELAND.   Curbing the spread of sand drifts was only one aspect of conquering the wastelands. Parallel to the movement of mastering the sands ran a more important program of land reclamation. In the Veluwe area, the moors were mostly either common or state-owned, but the question of ownership had become completely obscure in some cases. As elsewhere in Europe, a political movement started to turn the wastelands into oases of agricultural progress. In a process that strongly resembled the English enclosure movement, commons were partitioned and distributed among shareholders. A formidable obstruction to the reclamation of the land was a lack of dung. For that reason dividing the commons happened slowly, and at times only on paper. In the Veluwe area, lack of manure forced some early reclamations to be given up. Although some roads were improved and canals were dug, the problem would not disappear. Land reclamation ultimately took the form of afforestation with pines. Planting trees might help relieve the scarcity of fuel, yet one of the strongest incentives to planting trees stemmed from demand for pit props in British and Belgian coal mines. Between 1833, the time of the first cadastral survey, and 1870, the area of forest in the Netherlands increased significantly, from 1,690 square kilometers to 2,250 square kilometers; thereafter, the process decelerated.

AFFORESTATION IN PRACTICE.   The second example shows how afforestation could work. The nineteenth-century Dutch state became the heir of many commons, mostly heath, poor grazing grounds, and sand dunes. Such an area existed in the northwestern part of the Veluwe. These domains lay in between a large number of hamlets whose inhabitants used the commons as grazing ground for sheep. In the eyes of the authorities, they were, like the lakes, not effectively exploited. As some attempts at farming the commons in the early years of the nineteenth century failed, it became obvious that the soil was too poor to sustain agriculture or cattle. Afforestation provided a realistic option, for wood was in high demand in England and Belgium. After the Napoleonic wars the Dutch state lacked the money to initiate afforestation. Therefore, just as with the drainage projects, an ingenious financial mechanism was developed to start the process. In the 1830s, after a most welcome cadastral survey, the state claimed to be the owner of these lands and sought to sell them in order to stimulate cultivation of the wastelands. The municipalities claimed use rights over these lands, however, and refused their sale without financial compensation. State of-

TABLE 3
Afforestation on the Former State Domains
of the Veluwe (Netherlands), 1844–1914

Year	Total Area (ha.)	Suitable Area	Hectares	Increase (%)	% of Suitable Area Reclaimed	Pine Forests (%)	Other Tree Species	Arable Land
1843	24,148	—	?	—	—	—		
1854	24,248	12,776	2,630	—	21	61	6	32
1864	24,247	16,617	3,842	46	24	68	6	26
1874	25,196	16,617	5,016	44	30	72	6	22
1884	25,196	16,617	5,878	17	35	75	6	19
1894	25,196	15,651	6,220	6	40	76	6	18
1904	25,168	15,646	6,319	2	40	76	6	18
1914	25,131	15,613	6,814	8	44	75	6	19

ficials wanted a deal: they would transfer ownership of the commons for a given sum, including the arrears in the land tax, if the municipalities would promise, in turn, to jumpstart reclamation by selling the commons to private owners. Those new owners would then be exempt from the land tax. In this way, 250 square kilometers of land changed hands in 1843, of which more than 60 percent was, according to the state, fit for cultivation.

Table 3 shows the pace of afforestation after the commons were handed over to the municipalities.

In 1854 one-fifth of the area was reclaimed, in 1874 30 percent was reclaimed, and in 1894 only 40 percent of what was considered possible was reclaimed. By that time the process of afforestation had stagnated, just as in the rest of the Netherlands. More than half of the former state domains remained wasteland. The limit of what private entrepreneurship could accomplish on this infertile land had been reached.

As the opening text stated, these lands were especially fit for the planting of trees and not for agriculture. By the late nineteenth century about 75 percent of all the reclaimed wasteland was planted as dense pine forest aimed at agroforestry, about 6 percent was planted with other species of trees, and less than one-fifth was reclaimed as agricultural land.

As in the case of the lakes, the twentieth-century public increasingly appreciated this wasteland for recreation. In the densely populated country of the

Netherlands, the scarce, uninhabited moors proved to be excellent open terrain for army maneuvers. Therefore, the process of afforestation was never really completed. Today, the Dutch have rejected densely packed pine trees in favor of forestry that aims at open forests and a more natural growth of vegetation. The largest heath fields can still be found in the areas of the state domains.

Conclusions

In reviewing the destruction of landscape, the exhaustion of energy resources caused by peat digging and deforestation, and the attempts at political control, two conclusions can be drawn. The first conclusion is that, given its profitability, peat dredging could neither be stopped nor slowed down, despite heavy taxation. No perceptible degree of ecological restraint guided human action, and that is a sobering conclusion. From the perspective of, in today's terms, ecological restoration, the options were rather good. It proved possible, by means of a carrot-and-stick approach of forced investments (in the case of land drainage) and fiscal incentives (in the case of afforestation), to drain the lakes and populate the derelict heath and sand dunes with trees. It is necessary to warn against anachronism. The two related processes of lake drainage and afforestation were not "ecological" solutions in the modern sense but must be seen from the perspective of a patriotic ideology, whose central concern was to make the land as fruitful as possible in order to restore the power of the state.

Another conclusion is that ecological problems proved to a large extent solvable by properly directed government action. Without pressure from the state, the lakes and heath would perhaps have remained derelict landscapes. The historical irony reveals a sea change in environmental values: many Dutch people today are thankful that the state's project remained unfinished and that some lakes and heath fields are still there.

CASE STUDY: THE GERMAN GREEN PARTY

In February 2004, delegates from several European Green parties met in Rome in order to found the "European Greens." This new party was intended to meet the needs of an increasingly unified public space in the European Union. One of the driving forces behind the foundation of the European Greens was the French-German politician Daniel Cohn-Bendit, who started his political career in the tumultuous year of 1968. At that time a student leader known by the name of *"Dani le rouge"* (Red Danny), he presided over demonstrations at the

Sorbonne University in Paris as well as at the University of Frankfurt. There was another prominent German Green leader at the foundation ceremony, to-day's Minister of Foreign Affairs Joschka Fischer. Cohn-Bendit and Fischer are well acquainted, having inhabited the same commune in Frankfurt during the 1970s. They were then members of the famous nondogmatic, left-wing group called the *Spontis*, advocating "spontaneous" political action in their fight against capitalist domination. The German press recently questioned Fischer's biography when it leaked out that he participated deliberately in violent street fights against police forces during the 1970s.

The German Greens are highly important in the new European Green party, not only because their home country is the most populous in Europe but also because their national party is politically one of the most successful Green parties, in Europe if not in the world. The sketch presented above of leaders Cohn-Bendit and Fischer illuminates the Greens' long journey from being the so-called anti-party party to serving as a coalition partner in the German federal government.

The Work of a Generation

Founded in January 1980 in Karlsruhe, the West German Greens celebrated their first electoral success in 1982, when they gained 8 percent during the elections in the federal state of Hesse. One year later the Greens won seats in the national Parliament. In 1985 the first Green minister was sworn in when Social Democrats and Greens formed a coalition in Hesse, and the new minister, Joschka Fischer, attended the ceremony in sneakers. Green governance in Hesse, and in West Berlin in 1989, proved short-lived, but from 1990 onward the Greens became a natural partner for the Social Democrat Party, forming governments in the states of Brandenburg, Lower Saxony, and North Rhine–Westphalia. In 1998 the political success story of the Greens culminated in their formation of a national government together with the Social Democrats. The so-called secret chairman of the Greens, Joschka Fischer, became vice-chancellor and minister of foreign affairs. The German electorate confirmed this government in 2002.

However, the Greens remain a little party which typically gains around 10 percent of the vote. Indeed, the party experienced electoral crisis following the German reunification of 1990. Like the Social Democrats, the Greens had no political answer to the collapse of East Germany and instead started a notorious campaign with the motto, "Everyone talks about Germany, we talk about the weather" (alluding to the problem of global warming). The German electorate did not appreciate this aphorism and during the next four years the (former) West German Greens held no seats in Parliament. The West German and East German branches of the Greens united only after the elections. Because of a special stipulation in electoral law, the East German Greens held some seats in the *Bundestag*. Thereafter, many commentators predicted a quick end to the Green project, yet they underestimated the fact that the Greens were deeply rooted on several levels of the political system and that they were supported by loyal voters who, nevertheless, punished the unwillingness of the Green leaders to accept the historical facts of 1989–1990.

Since the founding of the Federal Republic of Germany in 1949, the Greens have been the only new party to have gained relatively stable support. How can we explain their outstanding success? There are several reasons. First, the Greens represent a political issue that emerged during the 1970s, namely environmental protection and the idea of an environmentally friendly lifestyle. The West German population paid much attention to the many concerns that fell under these categories. There was no other political party that concentrated on these problems with the same intensity as the Greens, and more voters were placing these environmental concerns at the top of their agendas. Around 1980, environmentalism was met by a second issue: peace. At that time, many West Germans did not accept the foreign policy of either Social Democratic Chancellor Helmut Schmidt (1974–1982) or his Christian Democratic successor Helmut Kohl (1982–1998). Both supported the NATO twin-track policy that included the deployment of new nuclear weapons in West Germany as a counterweight to Russian missiles on the other side of the Iron Curtain. A substantial percentage of Germans, mainly the young generation, did not want to have new atomic weapons in their country. They feared an escalation or, on the other hand, did not expect Soviet military intervention at all. Pacifism was popular. Although subsets of the Social Democrat Party did not agree with Schmidt and Kohl, the Greens were the only party that openly supported the goals of the peace movement.

One of the representatives of the West German peace movement was Petra Kelly. A charismatic person who exuded youth, modernity, pacifism, and self-assurance, Kelly was to become an important leading figure during the first decade of the Green party. She succeeded less in building a stable political network within the Greens than in influencing voters. From the late 1980s she lost most of her influence within the party, and in 1992 Kelly was shot and killed by her partner Gert Bastian, another Green politician who afterward committed suicide.

The peace movement provided a third basis for the Greens' success. The Green party was an organizational melting pot of several social movements that had emerged during the 1970s and that constituted a new political and social force. Some peculiarities of the West German (nowadays, German) political system encouraged their amalgamation, namely the practice of proportional representation, moderated by the 5 percent hurdle. The 5 percent hurdle stipulates that a party will be represented in Parliament only if it receives at least 5 percent of the total vote. This prevents the presence of splinter parties in Parliament. Another important feature allows the reimbursement of election costs with public funds. Germany's federal structure itself allows small parties to build regional strongholds without necessarily being successful on the national level. Finally, scientific research on the Green electorate suggests that the Green party is a kind of "generational project." It seems that the hard core of the party's supporters belongs to a single generation, roughly that of Daniel Cohn-Bendit and Joschka Fischer, born around 1940–1950. Green topics and the political style of the party's leaders appear to express the concerns of a certain group that soon will reach retirement age. (Klein and Falter 2003)

The Early Years: Grassroots Democracy and Ideological Schism

Let us bring the formation and early history of the German Greens into closer focus. The beginning of West German environmental policy can be dated to the year 1970, when the federal government announced an environmental protection program. Its official text stated that both natural and artificial environments required immediate measures to protect the health and safety of West German citizens. Rather than reacting to a public debate or the impact of a social movement, the government created a new field of political action, following the example of American environmental policy. Approximately at the same time, an international debate about the future of the planet emerged, focusing on overpopulation, resource scarcity, and the presence of toxic materials in human environments. The public increasingly became aware of apocalyptic visions of an allegedly imminent end to human life, or at least the breakdown of the existing economic system due to a lack of resources. These debates culminated in 1972 when the Report to the Club of Rome was published, a report that notably alarmed the West German public. At the same time government officials harshly criticized what they deemed "environmental hysteria." In fact, the government soon lost control over the environmental debate.

Additionally, new social forces, the so-called "citizen initiatives," began to discover environmental issues in the early 1970s. During the first years, primarily urban citizens founded local organizations to combat such problems as congested traffic, air pollution, and the lack of urban green spaces. Other groups intended to advise local or municipal administrations on environmental protection. Conflicts over nuclear power began to dominate the public debate on the environment around 1975. Several rural protest movements opposed the implantation of nuclear power plants in their regions. Not only did students or the well-educated, urban middle classes demonstrate to fight this polluting technology and the allegedly authoritarian "atomic state," but farmers did as well. Many of the existing urban environmental protection groups joined this debate and made nuclear power one of their main causes. Moreover, left-wing splinter groups joined the ranks of the environmentally concerned. Since the days of 1968 they had tried to turn the working class to revolution without success, but now farmers seemed to be willing to take a stand. Between 1975 and 1979, antinuclear demonstrations regularly attracted tens of thousands of participants.

In 1977 fights between the police and the opponents of nuclear energy showed on several occasions that protest alone would not stop officials from realizing the nuclear program. Public opinion remained favorable to the antinuclear movement, with about 50 percent of West Germans disapproving of atomic energy. Consequently, some antinuclear groups began to think about a new strategy: why not participate in local and regional elections? Politicization was problematic, however, for many members of citizen groups were convinced that party politics had resulted in a spoils system that no longer took into account people's real interests. They advocated loosely organized, nonprofessional initiatives as a democratic alternative to what they viewed as a corrupt political system. Although leaders of the environmental groups threatened to found "Green" voting lists if the conventional parties would not change their environmental policies, many environmentalists hesitated.

The antinuclear movement not only consisted of young or left-wing activists, for West Germany also possessed a long tradition of nature protection. Since around 1900 this bourgeois movement had developed steadily, relying on a conservative discourse that blamed modern urban society for alienating humankind from its roots in nature. Its representatives had close contacts with public services and administrations. By the 1960s, however, the conservation movement had become outdated. Its supporters belonged to an older generation that detested all kinds of western popular culture and advocated a rather authoritarian political model. Around 1970, however, some younger representatives of the conservation movement did not want to relinquish the environmental debate entirely to the citizen initiatives and the antinuclear movement. They began to advocate "ecology" as a model to describe the state of the environment. Within a couple of years a large portion of the environmental movement had adopted "political ecology" as a model for a better society. It included a vision not only of less pollution and protected nature but also of a less hierarchical society organized in small units. For political ecologists, the exploitation of nature was intimately linked to the exploitation of humanity. Political ecology was to become one of the cornerstones of the Green Party.

The first step in founding the Greens was taken by a rather conservative man, Herbert Gruhl. Gruhl was a member of the Christian Democratic Party and had specialized in environmental policy since around 1970. In 1975, he issued a best-selling book with an apocalyptic vision of humanity's future. (Gruhl 1975) His fears centered on overpopulation and future scarcities of energy and clean water. Like his allies in the environmental movement, he was convinced that the existing political system was not able to resolve environmental problems because politicians would only serve special interests. In contrast to the libertarian visions of the majority in the antinuclear movement, Gruhl advocated an "environmental dictatorship." In 1978 Gruhl left his old party and founded the "Green Action for the Future" (*Grüne Aktion Zukunft*), intending to run for deputy in the European Parliament in 1979. As his party remained insignificant, he joined forces with a rather conservative, middle-class splinter party named the AUD and two other conservative forces, the Green List for Environmental Protection in Lower Saxony (*GLU Niedersachsen*) and its counterpart in Schleswig-Holstein (*GLU Schleswig-Holstein*). These electoral lists had emerged from environmental groups that had participated in elections at the district level. Though they consisted of only a handful of members, they had been successful, the GLU receiving 3.9 percent of the votes in Lower Saxony's state elections in the summer of 1978.

Finally, all these groups participated in the European elections in June 1979 with a common platform. The result of 3.2 percent for the electoral coalition called *SPV Die Grünen* was not overwhelming, but it received 4.5 million German Marks (about 2.25 million euros) in reimbursed expenses. This money enabled the leaders of the SPV to organize a national party. In the autumn of 1979, Gruhl and his colleagues made the decision to launch a national Green Party on the basis of *SPV Die Grünen*.

Meanwhile, left-wing groups had rallied, too, in coalitions often called "colored" lists, such as the Hamburg list that was dominated by the Communist Union. The colored lists represented the young, urban, alternative milieu. Their members became aware that they would never be able to gain enough votes to jump the 5 percent hurdle, so many of these groups decided to join the conservative Green Party of Herbert Gruhl. They did so by the thousands in December 1979. By January the conservative and bourgeois founders of the Greens represented only a minority of the party's membership. The official foundation congress of the Greens (January 12–13, 1980) was dominated by a debate on whether members of the Green Party would be allowed to remain members of other parties as well, yet this problem pertained to communists only. Gruhl's concept of a traditional political party, which would be organized by a top-down structure, failed. He did not manage to become one of the party's spokespersons as he had wished, and he left the Greens in 1981. After some months he founded a new conservative ecological party, which never gained any importance. The Greens became politically strong because a variety of young and rather alternative groups rallied around the idea of environmental protection.

The Green Party was a melting pot. Its founding forces comprised conservative ecologists, antinuclear activists, anthroposophists, supporters of organic farming and an environmentally friendly lifestyle, the so-called ecosocialists who wanted to jettison capitalism in favor of an ecological utopia, and members of left-wing splinter groups. The Greens' political program was as heterogeneous as its supporters. Political scientists have called the Greens a frame party (*Rahmenpartei*), one that sets only a programmatic framework for its various constituencies without having a clear common program. Thus the Green Party functioned as a platform enabling its members to pursue different policies. Whereas Greens largely agreed on their political goals, they differed with respect to means.

Nevertheless, the Greens passed a political program in March 1980 called the Saarbrücken Program. It defined the character of the party as "ecological, socially minded, grass-roots democratic, and non-violent" (*ökologisch, sozial, basisdemokratisch, gewaltfrei*). These adjectives defined the political compromise that enabled the various groups to work together. Ecology was invoked as a result of the laws of nature: humanity could not continue to live in violation of natural rhythms. Environmental protection would have to dominate future policies. The new party was not able, however, to define how to achieve this aim. Some factions only wanted to reform the existing system by means of taxes and prohibitions. A second group advocated a new "third way" of economics, rejecting both capitalism and communism. Others wanted to prepare an ecosocialist revolution. The membership managed to agree on several demands, above all imminent and unconditional denuclearization.

The second principle of Green policy was social equality. On the one hand, this meant the expansion of the social security system. On the other hand, this slogan integrated a variety of demands that had been made by the new social movements: women's rights should be strengthened, abortion had to be legalized, discrimination against minorities had to be abolished, the conditions in prisons had to be improved, the exploitation of the Third World had to be stopped, and so on. The "nonviolence" of the Greens was also a legacy of the citizen initiatives. Their majority refrained from violent conflicts from around 1977, when communist groups hoped to unleash a revolt. The latter never really dominated the antinuclear movement, and by their strategy of violence they lost most of their influence within the movement. The sympathy expressed by much of the public for environmental protection groups flowed from the groups' nonviolent identity. Nonviolence symbolized the willingness of the Green Party to accept the rules of the political and legal system in West Germany, and it implied a second message that was related to foreign policy. The Greens agreed unanimously with the demands of the peace movement: Germany should ban all nuclear weapons, and NATO and the Warsaw Pact had to be dissolved.

The paradox of the early Greens was that they wanted to exert influence within the West German political system without becoming part of it. The Green Party was called an "anti-party party" that would not suffer from the corruption of ordinary politics. Many supporters and members of the Greens were not in favor of the "system." They continued to dream of an alternative, inspired by the political behavior of the citizen initiatives. Therefore, grassroots democracy (*Basisdemokratie*) became one of the most important elements of Green identity. Possibly it was grassroots democracy that guaranteed the cohesion of the Greens as much as environmental protection. What did grassroots democracy mean? During the first years it was inspired by the idea that the Green Party was only the parliamentary arm of the citizen movement. The party congresses of the Greens, for example, were open to nonmembers who spoke for citizens' groups. But the movement's involvement turned out to be short-lived; subsequently, grassroots democracy became a principle of inner organization.

Grassroots democracy had no elaborated political theory: it was inspired by a deeply rooted mistrust in political institutions, power, professional politicians, and all kinds of hierarchy. Paradoxically, the Greens turned their mistrust in the parliamentary system against the representatives of their own party. Nothing could be more dangerous, many of them thought, than a politician who lost contact with ordinary people. Grassroots democracy was an attempt to remain pure within an allegedly immoral political context. Thus the Greens set up a number of measures in order to ascertain that their representatives would remain in close relationship with the "simple" members of the party and the population. First of all, the Green Party initially had no paid positions. Second, no party official could run for Parliament, and no Green member of Parliament could have a post within the party's administration. Third, every member of Parliament had to resign his seat after half of the term of office to be replaced by the next in line (the "rota system," *Rotationsprinzip*). Green members of Parliament were obliged to turn a part of their parliamentary allowances to the party. Theoretically they had a fixed mandate, so their votes in Parliament had to follow exactly the decisions of the grassroots. In the beginning the Greens were convinced that party congresses could really render decisions on all important matters discussed in Parliament. Fourth, strict separation had to be maintained between the parliamentary parties and the party leadership. All important party offices were divided: for example, there was not one party leader but rather three (later two) "spokespersons" (*Parteisprecher*). All party offices and all seats in Parliament had to be equally divided between men and women. The German Greens were the first important social and political force to experiment with gender parity.

Grassroots democracy backfired in many ways. It could not prevent the development of internal oligarchies and informal networking. It changed to quite the opposite: as the checks and balances of the party structure prevented the accumulation of power by official means, the factions were obliged to organize informal networks, such that the decision-making processes turned out to be unclear. Moreover, competing tendencies within the party could easily block each other because of the structural rivalry between the parliamentary party and the party leadership. Although the Greens detested professional politicians, the logic of their grassroots system created politicians who "rotated" from a mandate to an administrative position, to another type of mandate, and so on. The prospect of a political career within the Green Party attracted people who had much time but no other important source of income. Joschka Fischer is a typical example of a Green functionary. He spent the 1970s planning revolution and working as a taxi driver and shopkeeper in a bookstore. In spite of his obvious intelligence he had no higher education. In 1982, a few days after the Greens scored an impressive success in the Hesse state elections, Fischer and much of his *Sponti* group joined the Green Party. One year later, Fischer took a seat in the national Parliament and became "manager" (*Geschäftsführer*, not speaker) of the parliamentary party. From this moment on he managed to remain a professional politician, switching among the national Parliament, the Hesse state Parliament, and his duties as State Minister of Environmental Protection for Hesse.

Grassroots politics turned out to be dysfunctional. During the 1980s Green members of Parliament often lacked the professional skills of their Liberal, Christian Democratic, or Social Democratic colleagues, whose parties allowed them enough time to gain experience in their respective fields of political activity. But grassroots democracy was the identification mark of the "anti-party party." Right into the early 1990s the majority of Green Party members were possessed by the idea that it was more important *how* they made politics than what kind of policy the party advocated. Party congress debates during the 1980s often concentrated on matters of internal democracy, whereas big politics was left more and more to the various party leaderships. The party's elite tried to reform the internal structures at the beginning of the 1990s, but the grass-roots were extremely reluctant to accept this. As late as 2002 a party congress refused the request of its spokespersons to take seats in Parliament. However, the famous "separation of party office and mandate" and gender parity remain the last bulwark of Green traditions, for most of the other grassroots principles have been abolished during the past decade.

The most important political cleavage during the first ten years of the Greens' history was the struggle between the so-called fundamentalists (*Fundis*) and the political realists (*Realos*). Soon after the party's foundation, disputes between these two factions replaced the initial conflicts between conservative and left-wing groups. As with the debate on the grassroots principle, *Fundis* and *Realos* did not discuss political programs as much as strategy. *Fundis*, on the one hand, wanted to maintain the Greens' identity as an anti-party party; they represented an unconditional opposition to the "ruling" system. The *Realos*, on the other hand, strived for (ecological) reform. One of the most important bulwarks of the *Realos* was the Hesse state branch of the Green Party. Several electoral successes during the 1980s confronted the Hesse Greens with the possibility of a coalition with the Social Democrats. In general, the southern state branches of Bavaria, Baden-Württemberg, and Hesse advocated *Realo* ideas, whereas the northern state branches of Berlin, Hamburg, and Schleswig-Holstein were dominated by *Fundis*.

At the end of the 1980s the internal quarrels threatened to paralyze the party. Conflicts grew more and more personal. The public witnessed party congresses that seemed to end in complete chaos. As this self-laceration did not appear to endanger the party's electoral success, there was little pressure to put an end to it. But after the disastrous 1990 elections, the Greens had to decide whether they wanted to participate in the realities of the new, reunified Germany or not. At this moment the majority turned against the *Fundis*, who left the Greens between 1990 and 1991. The thirteenth party congress declared in April 1991 that the Greens did not fight against capitalism, but advocated a policy of "ecological reform." Involvement in state and national governments became an official strategic goal. When East German and West German Greens merged in 1993, the unified party adopted a so-called consensus that was based on civil rights, democracy, and equal rights for men and women. For the first time, the Greens officially accepted the parliamentary system.

Only in 2002 did the Greens adopt a new and very different party program. The new program stands on the pillars of ecology, self-determination, equality, a "living democracy," and sustainability (the latter is one of the most important ideas in the repertoire of Green political concepts). The Greens not only demand a sustainable relationship between humankind and the environment but also extend the concept to economic and social policy. The new Greens advocate reductions in the national debt and public expenditure. Future generations must not be burdened with the costs of present-day policy, they argue, nor should consumer habits and the economic development of the industrialized nations restrain the opportunities of Third World countries. In comparison to that of the Social Democrats, Green economic and social policy has become "liberal" in the classic sense. During recent years, the Greens have appeared more eager to reform than their partners in government.

The Red-Green Government

In the course of their participation in the federal government since 1998, the Greens have had to face several problems. Only a few months after their coming into power, the Red-Green government (the Social Democrat/Green coalition) was confronted with the question of whether Germany should involve itself in military conflicts, first in the Balkans and later in Afghanistan. In March 1999 NATO started its attack on Yugoslavia and triggered the Kosovo War. Unified Germany was asked by its allies to fulfill its military duties as a middle-sized political power. Since World War II, German foreign policy had remained extremely cautious; successive governments designated the German military forces exclusively for national defense. West German foreign policy had to avoid the impression that Germany would return to its expansionist and militaristic historical traditions. With respect to international security policy, the "postwar era" began only after German reunification.

For their part, Greens had never had to take responsibility in foreign affairs, and pacifism remained one of the cornerstones of Green identity. The party had never revised this tradition and had failed to prepare for a situation, although it had become foreseeable after 1990, in which the Federal Republic of Germany would have to intervene in military conflicts. When Green cabinet members were confronted with war in the Balkans, they had no time to start a discussion in the party. The Green leaders accepted to send troops—not least because international pressure was intense and the Social Democrats were in favor of doing so. It was difficult for many Green Party members to accept this. If we take into account the Greens' grassroots tradition, it is nothing short of astonishing that the party neither obstructed this decision nor fell to pieces. However, emotional and heated confrontations dominated the Bielefeld party congress of the Greens in May 1999. Key factions advocated a resolution that would have forced the Green leadership to resign from government. Participants threw paint bombs at Joschka Fischer, who vigorously justified Green military policy. Some of his Green friends greeted him as "warmonger." In the end, the Biele-feld congress was a success for the Green members of government, who could continue their policy, but during subsequent months between 2,000 and 3,000 Greens left the party.

The Greens did succeed in one of their longstanding projects, denucleariza-tion. Although it was not possible to close down all nuclear power plants in Germany at once, the Green Minister of Environmental Protection, Jürgen Trit-tin, arrived at a compromise with the energy industry. In 2000 Germany be-came the first industrialized country to renounce atomic power; the last nuclear power plant will probably be cut off from the grid around 2040.

In proportion to their share of the vote, the Greens are not a big party. Be-tween 1987 and 1998 the party had 40,000–50,000 members in the western parts of Germany. In the eastern parts the Greens never surpassed 3,000 mem-bers. The members are well educated, and half of them have positions in the civil service. In 1998 the average age of party members was relatively low, compared to the other political parties in Germany. But whereas in the 1980s the Greens' membership and electorate had been dominated by very young people, they are now composed of the middle-aged. The Greens and their sup-porters have grown older as Fischer's generation continues to supply the core membership. Younger voters stand divided between the Greens and traditional parties, yet in their eyes the Greens now *are* a traditional party. Moreover, en-vironmental protection and peace do not dominate the political agenda of vot-ers in their twenties, who are more concerned about jobs, the economy, and so-cial equity.

The Greens not only play a role in national or state politics but also are en-gaged in local politics. Although the first Green mayor of a major city was elected only in 2002 (in Freiburg), there were around 6,000 local Green coun-cilors as early as 1988 in West Germany. In many cities the Greens challenged the traditional style of local politics. The latter had been characterized for decades by consensual decision-making structures and a narrow interpretation of the political function of local authorities, who would seldom meddle in mat-ters of national importance. Driven by their ecological worldview, the Greens made no distinction between local and national levels of environmental poli-tics. During the 1980s, as they remained excluded from the decision-making structures of the established parties, the Greens tried to politicize local issues as much as possible in order to launch public discussions. As a result, local poli-tics became more conflictual. But it was not only the Greens who caused these changes; elements of the Social Democrat and Liberal parties contributed, as did changing approaches in the local media. In the end, the Greens became integrated into local politics. They gave up extraparliamentary forms of political activity and concentrated on the town councils. The political style of the Green mayor of Freiburg has differed only very slightly from that of other mayors.

Significance of the Greens

The history of the Greens is a story of the successful integration of contesting forces into the German political system. It shows that the realities of the political system were stronger than the Green ideals of grassroots democracy. The system was attractive enough to motivate the Green leaders not to withdraw from it, and though several groups left the Green project during its history, often new supporters replaced them. Some leaders, like Fischer, discovered in the 1970s that radical left-wing politics did not succeed in West German society. After a long history of "normalization," the Greens ended up representing the mainstream of German middle-class values.

During the last quarter of the twentieth century, an "ecologizing" of (West) German political and everyday culture took place. While less than exact, this term highlights the impact of public environmental protection measures on the one hand and the willingness of the majority to behave in more "environmentally friendly" ways. State-sponsored programs for renewable energies have helped to increase the number of windmills and private households using solar technology. Business, too, has discovered the economic dimension of these developments since the 1980s, producing and marketing ecological consumer goods ranging from building materials to the famous *Bioladen* (food stores). Moreover, the public relations strategies of private enterprises often seek to create an ecologically friendly image: "eco" signifies high-quality products or services and the moral advantage of being kind to the environment.

One might conclude from this that the Greens have altered German politics and private life profoundly. We should be very careful, however, when weighing this interpretation. First, many other countries in the western world (with less vigorous Green parties) have undergone similar processes. Second, how might a small political party have had so much influence on everyday life and the economy? The opposite may just as well be the case: it is possible that the Greens could flourish *because* middle-class Germans, by the 1980s, had adopted a certain ecological consciousness and were apt to spend money for environmentally sound products. Third, Germany's environmental policies developed during the 1980s and 1990s, when the Greens were not part of the federal government. One of the milestones of German environmental policy in the early 1980s was the introduction of the catalytic converter in order to reduce air pollution caused by automobiles. Yet its introduction was spearheaded by the Christian Democrat/Liberal government of the time. It was during the 1980s, too, that the conservative government of Chancellor Helmut Kohl started to militate for high ecological standards within the European Union. Moreover, the traditions of German environmental legislation reach back into the 1960s and the early 1970s. Fourth, many environmental protection measures passed since the 1990s have stemmed from guidelines issued by the European Union. In many respects, European environmental policy has become a supranational agenda of the bureaucracy in Brussels. Fifth, whereas the Greens profited very much from their ecological image, at many times in their history their political program focused on other subjects, including disarmament policy, human rights, and immigration.

To sum up, the (West) German Greens have not caused the "greening" of German society, but they are merely an effect of it. Other political forces have made very important contributions to environmental protection. One should not be left with the impression, however, that the ecological paradigm has ever dominated German politics. It has always played second fiddle to the problems of the social security system and the desire for economic growth.

CASE STUDY: FROM NATURE CONSERVATION TO SUSTAINABLE DEVELOPMENT: THE SCANDINAVIAN EXPERIENCE

The Scandinavian countries—Sweden, Norway, Finland, Denmark, and Iceland—are densely populated, and, with the partial exception of highly agricultural Denmark, they are famous for their wilderness and extensive areas of unspoiled nature. Historians started to write about people and nature in Scandinavia in the late 1800s, giving us environmental perspectives on agriculture, forests, water, and landscape. Over the past two decades the environmental consequences of industrialization have received much scholarly attention, as have the political challenges in meeting what we speak of today as sustainable development. Internationally the Scandinavian countries, and Sweden and Norway in particular, have received much publicity for promoting progressive environmental politics. Sweden's reputation took off with its sponsorship of the United Nations Conference on the Human Environment in 1972, and Norway shone in 1987 when a United Nations commission led by the former Norwegian prime minister Gro Harlem Brundtland published *Our Common Future.* Both events will be treated below. The recent international role played by Sweden and Norway justifies the focus on these two countries in much of this case study. Historical scholarship is also weighted toward nature conservation in Sweden and Norway. This case study covers the main lines of the Scandinavian experiment regarding the origins of nature conservation, how this experiment was carried out, and what changes took place when sustainable development became the guideline for nature conservation and environmental protection.

Scandinavian Conservation and the German Model

A desire to protect nature can be traced back to the early 1900s, and many natural scientists conjoined to produce ideas about nature conservation. Scandinavian scientists found inspiration from pioneers like Gifford Pinchot (1856–1946) and John Muir (1838–1914) in the United States, but even more directly from German professor of botany Hugo Conwentz (1855–1922). He traveled all over Germany, giving lectures and speeches to government officials and academics, and ultimately became the foremost figure in state-sponsored nature conservation in Germany. In the years between 1900 and World War I he wrote several books and brochures about nature conservation. He was well known among Scandinavian natural historians, and during 1904 and 1905 he also traveled in Scandinavia, giving lectures on nature conservation. Conwentz and the German nature conservation campaign paved the way for the Scandinavian model of nature conservation in the early 1900s. This model linked nature conservation to developments within the natural sciences, to the creation of the welfare state, and to the shaping of nations.

Conwentz viewed nature conservation through a cultural lens, seeing it as an instrument for nation-building. His ideas spread during the Second German Empire (1871–1918), a period that saw structural changes in society in tandem with accelerating industrialization, technological development, and deteriorating ecosystems. Of crucial importance was how the nation should be built in the years to come. Conwentz grounded his ideas on nature conservation in a critique of modernity focusing on the negative effects of rapid industrialization. He voiced particular concern over damage to forests and the countryside caused by the construction of railroads and highways. Conwentz also criticized the dredging and canalizing of rivers, the building of dams and electric power lines, and pollution of air and water. For Conwentz, nature conservation was a means of shaping the nation along other lines. He stressed the importance of imparting knowledge of the natural world to younger generations. Modern Germany should be a nation of scenic beauty, with a healthy countryside that strengthened German national identity.

Conwentz circulated his ideas during a time when voluntary societies and associations of many kinds flourished in Germany. Through these structures, German conservationists presented several arguments for their cause. First, they argued, nature should be protected for recreational reasons; second, endangered species should be preserved for moral as well as scientific reasons; third, the spoiled countryside undermined patriotic loyalty to the German nation; fourth, people's health was in danger due to pollution; and fifth, the present generation had moral obligations to future generations.

The significant rise of the nature conservation movement in Germany up to World War II resulted in both governmental and private activities. New institutions linked to conservation were the Prussian Governmental Center for the Preservation of Nature, the Nature Park Society, and the League for Nature Conservation in Bavaria. Among the pioneers were primarily scientists and representatives of the new and growing middle class.

Among the Scandinavian countries, Sweden was a pioneer both in terms of the institutionalization of nature conservation and its content. The Swedish Association for Nature Conservation was established in 1908 in order to "preserve the love of Swedish nature and conserve it." The statement implied that human activity and particularly industrialization threatened the natural environment, and the association's members were to encourage all Swedes to be stewards of Swedish nature.

One of the most important debates animating nature conservation centered on the creation of national parks. In this area Sweden did not follow the prescriptions of Conwentz. Unlike pioneers in many other countries, Conwentz did not recommend the designation of national parks. He favored protecting particular scenic spots and communities of rare plants. The Swedish government, however, decided in 1909 to create nine national parks, and this made Sweden unique in Scandinavia. Six of the parks, including the largest, were located in the northern parts of Sweden. The Swedish government regarded the North as backward and disconnected from the rest of Sweden. As Sweden industrialized in the second half of the nineteenth century, authorities sought to integrate the North into the nation. This was Sweden's frontier, and much like the American West, the Swedish North came to symbolize modernization itself. On the one hand, the North was a productive area with natural resources to be exploited, namely large forests and rivers. On the other hand, Swedish politicians and scientists also pursued modernization by mobilizing the population around the aesthetic and cultural values of nature in order to shape a strong Swedish identity. If Swedes learned to love their own natural heritage, they might also be dissuaded from emigrating. Finally, government officials assumed that the North, once "put on the map" with its national parks, would attract tourists.

In Norway leading conservationists strongly supported Conwentz's ideas about nature conservation. However, other groups had different opinions regarding conservation models. Norwegian professors of history, geography, and botany worried about "old Norwegian nature," that is, the rivers, forests, and untouched mountains that needed protection. They suggested, with reference to Yellowstone National Park in the United States, that Norway should create national parks. The Department of Agriculture investigated the possibilities for creating a national park in southeastern Norway, close to the Swedish border, mainly because of the large forests and the particular flora in this area. The plan was shelved, and Norwegians had to wait several decades before the country got a national park. Politicians of all parties claimed that Norway had "enough" nature, and public policy remained geared toward economic growth. Norwegian authorities did, however, pursue conservation along the lines of Conwentz's philosophy of protecting small, endangered plant communities.

Even if the quest for national parks proceeded haltingly in the early 1900s, nature conservation was put on the political agenda due to the negative consequences of cruise-ship tourism along the coast of northern Norway. Passengers on German cruise ships had the habit of painting the name of their ship on coastal rock walls. One of the most serious incidents happened when the name of the German emperor's ship, *Hohenzollern*, was painted on the North Cape Rock. Beyond deploring what appeared to be an aggressive political message, many Norwegian conservationists viewed the act as serious defacing of the landscape and asked the Ministry of Agriculture to intervene. At first the minister did not see the painting as a problem, retorting that the North Cape Rock was "ugly and naked and deserved to be decorated." Conservationists saw it differently and wanted to stop the paintings. They made Conwentz aware of the defaced rock, and he informed the German authorities. Soon the habit of painting ships' names stopped, and the Norwegian authorities also changed their attitude, removing all the paintings in 1909.

Following these events in the northern part of Norway, scientists advocated new nature conservation initiatives. Like their German and Swedish counterparts they viewed nature conservation as part of the construction of a nation's culture and civilization. They were concerned about population growth and human activity that caused changes in the landscape. The breakthrough came in 1910 when the government signed the Nature Conservation Act. The act stated that the king could decide to conserve particular areas when necessary for the protection of flora and fauna of particular scientific and historical importance. For the first time in Norwegian history, nature conservation had become a political issue. In the following years the government sought to protect areas of particular botanical and historical value. Otherwise, nature conservation was organized on a voluntary basis, and the first association for nature conservation in Norway was established in 1914. Typical of its time, the organization consisted largely of male professors, often with backgrounds in the natural sciences, as well as state officials with backgrounds in law.

The young nature conservation movement rapidly met challenges. Industrialization swept the Scandinavian countries, and nature conservation arrived in the midst of political debates about modernization and tradition. In a mountainous country like Norway, with its untamed waterfalls, conflicts between 1905 and 1917 swirled around concession laws. Debate centered on whether or not foreign investors should have the right to purchase Norwegian waterfalls for hydroelectric development. The Conservatives favored technological development, as did the Social Democrats, while the Liberal Left feared the loss of Norwegian wilderness to foreign capital. Concession laws ultimately allowed public purchase of the rights to waterfalls, but they made no mention of nature conservation.

As has often happened, nature conservation was initially the loser in the battle between technology and nature. Norway lost several waterfalls in the first half of the twentieth century. One of them, Skjeggedalsfossen, had been one of the most spectacular waterfalls in northern Europe and a major tourist attraction for decades. When the Norwegian Tourist Association celebrated its fiftieth anniversary in 1918, it promised to conserve the waterfall for the future. The promise was put to the test the very next year. A new industrial center had developed close to Skjeggedalsfossen, with a company that established the electrometallurgy industry in the region, and subsequently the company needed more hydropower. The Norwegian Tourist Association realized the dilemma, but its board argued that the waterfall's attractiveness to tourists had declined dramatically since industry had moved in.

Lively debate circulated among members of the Norwegian Tourist Association (which held the easement to the surrounding land), local landowners, and politicians. The tug of war continued for more than ten years, ceasing in 1932 when the Supreme Court decided that the Norwegian Tourist Association could sell the easement to industry. The association's argument that the waterfall was already spoiled as a tourist attraction seemed to prevail, but even more persuasive was the electrometallurgy industry's creation of jobs in a period of economic crisis.

In Denmark discussions about nature conservation were reflected in the press in the latter half of the nineteenth century. One of the first burning issues concerned the loss of scenic views caused by natural afforestation. Threatened visual perspectives around sites of recognized aesthetic and historical value, such as the Castle of Frederiksberg, sparked debates in conservation circles for over thirty years, with little result until shortly after 1900.

The year 1905 saw a turning point in the case for nature conservation in Denmark. That year scientists from the Natural History Association, the Botanical Association, and the Geological Association invited Conwentz to lecture in Copenhagen. His lecture and the case for nature conservation received much attention in the politically conservative newspaper *Berlingske Tidende*, whose writers stressed the importance of protecting the open landscape, under pressure from urbanization. According to *Berlingske Tidende*, however, conservation should not be a hindrance to industry or agriculture. Scientists argued along similar lines and thus also found a voice in *Berlingske Tidende*. The Tourist Association at Skjælland spearheaded the founding of the Nature Conservation Society in 1911, and the first law on nature conservation came in 1917. As in Norway, the Danish government supported Conwentz's model and focused on preserving particular landscapes close to cities. Officials justified conservation on scientific and aesthetic grounds and also spoke for the rights of future generations to experience Denmark's "original" nature.

During these same years Danish radicals adopted and redefined nature conservation through their newspaper *Politiken*. They fought for public access to the coastline and to forests and thereby placed the question of access at the center of conservation. Promoting access to coastlines was, for example, inextricable from halting the development of coastal zones by wealthy people intent on fencing their properties.

While Sweden, Norway, and Denmark were influenced by German conservationism, Finland's road to nature conservation followed a different path. One important initiative came from Finnish professor of history Ernst G. Palmén. In 1903 he wrote an article criticizing the lowering of lakes as an environmental problem. The desire to preserve landscapes also contributed to forming Finnish nature conservation, for the Finish people were highly aware that their forests and waters formed the basis for a national mythology. For historian Timo Myllyntaus, wilderness could, in effect, "stand for the strength and determination of the Finnish people in its struggle against russification and in dreaming of political independence." (Myllyntaus 2004, 2) Shortly after having gained independence in 1917, Finland was wracked by civil war and economic depression. The Finnish government placed environmental issues on hold until 1923 when it passed an initial law on conservation. As in neighboring Scandinavian countries, however, the interwar depression halted further initiatives.

Depression, War, and Affluence

During the interwar period, with economic depression and mass unemployment, nature conservation throughout northern European countries reflected not only scientific research but also the rising tide of nationalism. In Germany the ideology and demands of National Socialism came to overlap with nature conservation. The famous "blood and soil" conservation slogan of the Nazis suggested that interaction with the soil of the Fatherland had shaped the German people's identity. Certain "German" qualities such as courage, loyalty, steadfastness, and honesty could not, therefore, be shared by non-Germanic peoples. (See chapter 5 for further detail.)

A tendency toward ideological overlap between National Socialism and nature conservation was also present in Scandinavia. In Denmark an extensive debate over the meaning of conservation attracted the high political and cultural elite. Radicals once again claimed the movement was reactionary and antimodernist; according to the radicals' viewpoint, the movement tasted of German fascism highly influenced by the Romantic tradition and thoroughly rejected Enlightenment values. Radicals noted that conservation targeted the welfare of the peasantry while ignoring urban populations. They accused it of favoring an obsolete, folkloric culture that tied the Danish people to the Danish soil.

In practice, conservation remained an elitist enterprise directed by scientists rooted in a conservative ethic. One of the leading voluntary organizations in Norway, the National Association for Nature Conservation, lobbied unsuccessfully to amend the Nature Conservation Act, in hopes of extending its scope in order to permit the creation of national parks. However, the organization did enlarge the concept of nature conservation internally, enacting reforms in 1936. Although the leader of the association was an active Nazi, Norway, in contrast to Denmark, did not host a national debate about the links between conservation and fascism. With the outbreak of World War II all nature conservation work, governmental as well as voluntary, stopped.

Europe after World War II saw rapid growth of heavy industry and the economy in general. Governments did not show particular concern for environmental effects, and conservation movements in many countries had not yet been restored. In the 1950s a few Scandinavians raised their voices, for instance the Swedish natural scientist Georg Borgström. Of main concern to Borgström was overpopulation, and he was highly skeptical of contemporary beliefs in infinite technological and economic development. In Norway the notion of a coming ecological crisis gradually gained attention. Of main concern to Norwegian conservationists after 1945 was the harnessing of rivers to provide energy for ironworks and electrometallurgy. Another issue was the decline in fisheries that followed from the advanced fishing technology of the 1950s. (See chapter 6 for more detail on Scandinavian fisheries.) One highly controversial topic was whaling. Norwegians had engaged in whaling for many decades and contributed to a near-extinction of the blue whale. The catch of whale had intensified in the early 1930s, and after World War II the quota was drastically reduced. The Norwegian government signed the International Convention for Regulation of Whaling in 1946. Despite reduced quotas, this agreement did not in itself revitalize whale populations. Likewise, not all nations respected the rules of the International Whaling Committee (IWC), established in 1949. When the committee met in 1963, the extinction of the blue whale was so close at hand that the committee agreed to total protection, which took effect for blue whales in 1965.

Nature conservation was not fully revitalized until the early 1960s. More accurately, the early 1960s saw a paradigm shift from nature conservation to environmental protection. Environmental policies, environmental movements, and a generalized environmental awareness began to emerge across the political landscape of northern Europe. The concept of "environment" hailed from the field of psychology and other social sciences, but it was now introduced to describe the ongoing debates about the relationships among humans, nature, and health. In northern Europe the notion of environmental protection now subsumed both nature conservation and pollution control. In practice Scandinavians still spoke of "nature conservation" when referring to its classical preoccupation with the protection of species and habitats.

In Scandinavia, as in other European countries and the United States, the new awareness and conceptual changes happened in tandem with scientists' warnings of a coming ecological crisis. In particular, Rachel Carson's book *Silent Spring* (1962) triggered the debate on environmental protection in Scandinavian countries. Carson located the origin of the most serious environmental problems within the agrochemical industries. Her message was that indiscriminate use of DDT and other insecticides would kill birds and other wild populations, for insecticides accumulate in food chains. In the United States she was accused of being hysterical and unscientific, particularly within the U.S. Department of Agriculture and among the scientific establishment, but her book was a public success. Carson had two prominent Swedish counterparts, Hans Palmstierna and Rolf Edberg, environmental actors and authors with roots in the Social Democratic Party.

Palmstierna approached environmental problems from a natural scientific and anthropocentric point of view. His main publication, *Plundring, svält, förgiftning* (*Plundering, Starvation, and Poisoning*), received much attention upon its publication in 1967. Palmstierna stood among other apocalyptic environmentalists and scientists of his time, for whom modern industrialized society faced overpopulation and pollution that threatened the survival of humankind. In order to resolve the crisis, Palmstierna advocated stronger institutional controls. His faith in Sweden's existing social democratic political system led him to argue that the state could best handle environmental problems through legislation and administrative control systems. With environmental protection in view, technology could be a prime instrument at the disposal of the state, but only if government and scientists worked more closely together. Not surprisingly, Palmstierna's ideas fit well with the ideology of the ruling Social Democratic Party, which sought to develop the Swedish welfare state based on industrial growth. Leading politicians were aware of the predictable environmental effects of their policies, but since Palmstierna also represented the social democratic ideas of planning and control, his proposed solutions to environmental problems did not threaten the political establishment. When the Swedish Social Democratic party drafted an environmental program in 1968, it was primarily informed by Palmstierna's strategies. One important outcome of the environmental debate in the first half of the 1960s was that in 1967 Sweden became the first Scandinavian country to establish a national environmental regulatory agency, the Swedish Environmental Protection Agency (SEPA). Two years later Sweden passed a comprehensive Environmental Protection Act that regulated air, noise, and water pollution.

Rolf Edberg was a journalist, longtime member of Parliament, and for ten years the Swedish ambassador to Norway. His first and probably most influential book was *Spillran av et moln* (*A Shred of a Cloud*) (1966). In contrast to Palmstierna, Edberg expressed a biocentric point of view, criticizing the technological and economic underpinnings of modern civilization. Far from agreeing with the current social democratic faith in new technology and continued economic growth, Edberg argued, with much pessimism, that the most fundamental changes in human behavior would be required to avert ecological catastrophe. And, he said, people would never learn to adapt to the laws of nature without experiencing it unspoiled. Edberg also took a more international approach than did Palmstierna. He wrote of the possibility of a global governance system that could halt population growth and reduce human pressures on nature. His ideas received, as we will see, much more attention among Swedish politicians some years later.

In Norway, environmental protection arose out of strong pressure from nongovernmental organizations (NGOs) and the reconstructed National Association for Nature Conservation. The society had lobbied energetically in the 1950s to revise the nature conservation law, and through the new law of 1954 the Norwegian government opted to create national parks. Still, no national parks were actually designated until 1962, largely due to the "sacred cow" of river development, supported by Norwegian Social Democrats to stimulate economic growth.

Norway's first national park was Rondane, a mountainous area with peaks higher than 2,000 meters and home to a herd of wild reindeer. Following the creation of Rondane National Park, discussion within government centered on a national plan to protect large areas according to the national park model. The National Board of Nature argued that Norway was one of the few countries in Europe where "untouched" nature still existed, referring to the country's still-uncultivated, ungrazed stretches. Thus the board advocated protecting several extensive areas, not only in the name of science, education, and the interests of future generations but also to provide an increasingly urbanized population with places for outdoor recreation.

Classical nature conservation issues were prominent in the Norwegian debate in the 1960s, but pollution and the notion of ecological crisis gained wider attention. One important contribution to this turn was Rolf Edberg's book *Spillran av et moln*, and in the late 1960s several of Georg Borgström's books were translated into Norwegian. A few years later a Norwegian philosopher entered the environmental stage. Arne Næss represented the Deep Ecology movement and took a biocentric approach to environmental issues. Two early and influential writings were his book *Økologi, samfunn og livsstil (Ecology, Community, and Lifestyle)* and his article "The Shallow and the Deep, Long Range Ecology Movement: A Summary," both published in 1973. Næss's ideas inspired NGOs, especially environmental groups fighting to protect Norwegian rivers and waterfalls. He was himself an environmental activist and was forcibly removed by police from a demonstration protesting against the regulation of one of Norway's most spectacular waterfalls, Mardøla.

Launching Sustainable Development

It is often said that the year 1962, with the release of *Silent Spring*, represented a breakthrough in modern environmentalism. Such was also the case with the year 1972. In that year, upon a Swedish initiative, the United Nations Conference on the Human Environment took place in Stockholm. The conference represented a new turn in global environmental policy and identified Sweden as an environmental pioneer. After the 1972 conference and also the publication of *Limits to Growth* by the Club of Rome, the environmental debate in Sweden intensified just as the oil crisis began and the long trend of economic growth ground to a halt. The strategies of Palmstierna and the Social Democrats of the 1960s no longer provided obvious solutions. Swedish environmental NGOs now became more active and critical of hopes that economic growth would cure all ills.

The debate took on broader dimensions between 1973 and 1975, focusing at first on energy due to the expansion of nuclear power in Sweden. Palmstierna was once again in the forefront, yet his earlier faith in economic growth had faded; he now argued in favor of a society based on low energy consumption. Despite a mounting opposition to nuclear power in Sweden, the Social Democratic Party supported its development, and its leaders fell from power in the election of 1975. During its time in opposition the party appeared to adopt an ecological orientation to environmental problems. Back in power in 1981, the Social Democrats proposed an environmental program centered on the protection of air and water, chemical control, recycling, and nature protection. The program stated for the first time that nature protection was a value in itself, indicating that Rolf Edberg's ideas, formerly neglected, had now become part of Social Democratic environmental policy. The 1981 program also had an international focus that anticipated Swedish policies following from the Rio conference (see below) eleven years later.

In Norway, too, environmental issues rose high on political agendas in the early 1970s. Norway became the first Scandinavian country to establish a Ministry of Environment. The public environmental debate was particularly linked to the discussion of Norway's entrance into the European Community. The ecological perspectives formulated by Arne Næss, along with environmentalists Hartvig Sætra and Sigmund Kvaløy Sætreng, dominated the opposition to membership. Claiming that Norway had a variety of environments, livelihoods, and rich natural resources already under pressure from industry and urbanization, they feared a loss of national power over their control if Norway were to join the European Community. The majority of the Norwegian people voted against joining the EC when the referendum was held in 1972.

Another crucial topic within the environmental debate was, as in Sweden, energy. Norway's Labor Party favored nuclear power but was met with massive protests from the environmental movement. Then, in 1979, the nuclear power accident at Three Mile Island near Harrisburg, Pennsylvania, contributed to the end of the debate, and to this day Norway generates no nuclear power. The debate over hydropower, however, continued. One of the most contentious debates related to the exploitation of the Alta-Kautokeino River in Finnmark. The environmental movement practiced civil disobedience and managed for a while to stop the construction work, which was then in an early phase. Following police actions and the removal of demonstrators, the government decided to build a hydroelectric power plant at the site.

Both in Scandinavia and beyond, environmental thinkers and policymakers linked environmental issues to economic development in new ways during the 1980s: the now common phrase "sustainable development" was coined. The concept of sustainable development figured first in the World Conservation Strategy report from 1980, yet it did not come into its own until the release of *Our Common Future* (1987), the report from the World Commission on Environment and Development (WCED). The United Nations General Assembly formed this commission in 1983, naming the former Prime Minister of Norway, Gro Harlem Brundtland, as chairperson. The commission framed sustainable development as a novel set of ideas and goals with respect to both nature conservation and environmental protection. We might ask, then, did the new conceptual framework change the way of handling environmental issues in Scandinavia?

It is possible to identify some changes, which are rooted in propositions from the WCED. First, the commission suggested long-term environmental strategies to be carried out in all nations. Second, it asked for better cooperation among countries of the northern and southern hemispheres in environmental matters. Third, it advocated more efficient international cooperation and common goals in order to solve environmental problems. By providing these guidelines, the commission saw the environment as part of a triangle consisting of ecological concern, economic concern, and social concern—all constituting the three pillars of sustainable development. The commission further required action to reorient economic growth and called for a type of development that integrated production with nature conservation. The notion of development was applied not only to developing countries but also to industrialized countries. Thus, sustainable development could be understood as a challenge common to all nations.

The Scandinavian countries have been regarded as forerunners in the field of environmental policy, or they have at least imposed the pioneering role upon themselves. Sweden, in particular, has been viewed as a model society when it comes to both welfare and environmental goals and laws. Sweden did indeed respond to the WCED and the United Nations Conference on Environment and Development (UNCED), held in Rio de Janeiro in 1992, in unique ways compared to other countries in northern Europe. One of the first important moves toward sustainability was the proposal from the coalition government (consisting of the Moderate, Center, Christian Democrat, and Liberal parties) in 1991 to draw up a new Environmental Code. This code, introduced in 1999, codified environmental legislation; it makes Sweden one of the most advanced countries in northern Europe within this field. Since becoming a member of the European Union, however, Sweden has been under pressure to weaken its environmental regulations and laws, though its representatives lobby for stricter rules within the EU.

When the Social Democrats governed again after 1994, Prime Minister Göran Persson declared in 1996 that Sweden should serve as an international model for the creation of an environmentally sustainable society. Of core importance was the linkage of sustainability to welfare, to the limitation of greenhouse gasses, and to the protection of biological diversity. Since then Sweden has also worked hard to set out national political strategies to follow up on the UNCED. On the innovative side has been Sweden's response to the UNCED at the municipal level. The discussions in Rio and the recommendation in Agenda 21 called for the delegation of environmental responsibility from national governments to municipalities and nongovernmental organizations (NGOs). Three years after Rio, more than half of Swedish municipalities had employed so-called LA21 coordinators. They organized information campaigns about Agenda 21, gave seminars or courses, and launched such projects as free busses or environmentally friendly tourism. Many municipalities soon introduced a range of activities relating to environmental education and new forms of participation. Major fiscal innovations have helped promote energy efficiency and increased use of renewable recourses, in line with efforts to create new jobs.

Sustainable development policy differs from pure environmental policy in that it contains the idea of sectoral integration. In this respect, too, Sweden has pioneered, as with green accounting in both the private sector and public agencies, and the integration of environmental criteria in allocating foreign aid. Several industries, including the large paper pulp industry, have implemented environmentally friendly management practices.

With the release of *Our Common Future* and the role of Gro Harlem Brundtland, Norway also took a seat in the international firmament. The government declared sustainable development to be the overriding objective of Norway's future development as a welfare state. One substantial result of the new orientations inspired by *Our Common Future* was an additional paragraph in the Norwegian Constitution. The amendment stipulated that every citizen had the right to an environment that did not threaten people's health, and the right to a productive nature, implying that natural resources should be managed for the long term, with regard for the rights of future generations. In the following years the definition of sustainable development and references to *Our Common Future* found their way into innumerable Norwegian governmental documents and publications. National efforts to follow through on the program outlined in *Our Common Future* and the Rio Conference's Agenda 21 represented both continuity with and change from earlier environmental protection work. Classical nature conservation issues remained prominent, for instance, the preservation of forests, mountains, and rivers. Pollution also received much attention, but was primarily treated as an end-of-the-pipeline problem. Thematic newcomers, however, were climate change and sustainable production and consumption. Norway led internationally in establishing an Interministerial Climate Group and the Environmental Tax Commission, which promoted heavier taxes on energy-intensive and polluting activities while decreasing employers' taxes on labor in order to combat unemployment. Internationally, Norway initiated discussion of sustainable production and consumption within the United Nations' Committee on Sustainable Development (CSD), sponsoring no fewer than three conferences to inform the CSD's work on Agenda 21.

Unlike Sweden and Norway, Finland and Denmark have not acquired international reputations for promoting sustainable development over the past ten to fifteen years, if measured by the hosting of conferences or forming prominent persons in this field. This does not mean that the two countries have not taken action at home, however. In Finland, the first step in promoting sustainable development was taken in 1990 when the government issued a report titled "Sustainable Development and Finland." Three years later it set basic national goals, introduced measures for implementation, and established the National Commission on Sustainable Development, a commission chaired by the prime minister. As in Sweden and Norway, the Finns have focused on sustainable development at the local level. The LA21 process has been crucial in Finland, as in the other Scandinavian countries, and in order to stimulate LA21 work, the environmental administration underwent reforms in 1995. Regional authorities acquired more responsibility in supporting municipalities seeking to integrate environmental concerns. More than 60 percent of all municipalities have initiated an LA21 process, but of course the level of activity varies. The process is continually in flux, and one main challenge is to unify the Finnish government's ambition to take a stance on environmental issues of global import, along with integrating citizens' desires to improve their local neighborhoods.

Soon after the release of *Our Common Future* in 1988, the Danish Ministry of Environment launched the "Our Common Future" campaign and an action plan "for Environment and Development." As in the Finnish case, the government encouraged citizens to engage in environmental matters and stimulated cross-sectoral thinking at the local level. A "Green Municipality Scheme," launched in 1988 by the Ministry of Environment and Energy, strengthened local efforts. Several green projects emerged, such as educating children on environmental issues, nature conservation, and cleaner technologies. Acting on early results, the Ministry formed a more comprehensive strategy between 1988 and 1992, entitled "Denmark Heading for Year 2018." The strategy aimed to give Denmark the role of environmental pioneer and forerunner in the promotion of sustainable development. The Ministry's efforts focused largely on spatial planning and urban development, which seemed logical since Denmark is the Scandinavian country with hardly any wilderness to protect. In Denmark, too, LA21 projects took on health and the quality of life, renewable energy, reduction of consumption, forest regeneration, protection of watersheds and drinking water, and education; now over 70 percent of Danish municipalities have initiated LA21 projects. One of the main results has also been a democratization of local environmental policies, from the definition of problems to the search for new solutions.

Even though Denmark's road to sustainable development has wended through local initiatives, there is one sector where Denmark should be mentioned as an international pioneer, and that is in the promotion of sustainable energy. In Scandinavia Denmark is a leader in developing wind power, and the country aims to become the global "Wind Power Hub." At home, wind power constituted 15.9 percent of total energy consumption in 2003. That year there was less wind than normal, and under normal conditions the proportion would have been 19.7 percent. Danish authorities expect that wind power will account for 25 percent of all energy consumption in 2008. Denmark exports wind power technologies to other countries, particularly to Europe and the United States, and the market in Asia is growing. Iceland has also led in promoting sustainable development through renewable energy. The Icelandic government cooperates with national and international energy companies to engineer a transition from fossil fuels to hydrogen. Iceland has been a forerunner at home, and today 90 percent of all energy comes from geothermal and hydrogen sources.

The affluent countries of Scandinavia have been highly supportive of the ideas and goals of sustainable development, and they have charted progress in important areas such as waste management, pollution control, and nature conservation. But consumer society, there as elsewhere, has yet to reduce its consumption of energy and nonrenewable resources. In practice, sustainable development has been furthered by a top-down, policy-oriented approach rather than a bottom-up, participatory approach. This fact has led researchers within the field to describe Sweden's efforts to create a sustainable society as "progression despite recession" (Eckerberg 2000, 209) and to depict Norway as a country "reluctantly carrying the torch" (Lafferty and Meadowcroft 2000, 174).

Despite these contradictions, Sweden and Norway remain in the forefront when taking an international perspective on sustainable development. There are historical reasons behind the enthusiasm in these two countries. Both of them have an open and democratic political culture and also a very active civil society. This culture paved the way for broad-based conservation movements 100 years ago and eventually for environmental movements that have achieved high political influence since the 1970s. Thus, prior to the era of sustainable development, we can identify a burgeoning environmental consciousness and well-organized environmental institutions, such as environmental ministries and national pollution agencies. Historically flexible administrative practices allowed for relatively easy integration of sustainable development and its cross-sectoral ideas. In the consumer societies of northern Europe the idea of sustainable development as a guiding principle has come to stay. The main challenge is to find implementation measures that manage to integrate environmental, economic, and social concerns.

CHRONOLOGY

Abbreviations:

YA: years ago
BCE: before common era
CE: common era

480 million YA: Europe and North America collide.

245 million YA: Supercontinent Pangaea takes shape in a collision with Asia. The Ural mountains are formed.

200 million YA: Pangaea begins to break apart.

80 million YA: Antarctica moves to the South Pole; accumulating ice initiates a long-term cooling trend.

60 million YA: North America begins to separate from Eurasia. Africa bumps up against Eurasia, creating the Alpine chain.

2.5 million YA: Beginning of the Paleolithic in Africa.

2.3 million YA: Beginning of the Pleistocene epoch, an ice age affecting mid-latitude areas in a series of glacials and interglacials.

1.8 million YA: Presence of hominids in the Caucasus.

900,000 YA: Presence of hominids at Gran Dolina, Spain (oldest human remains found in western Europe).

690,000–550,000 YA: Evolutionary lines leading to *Homo neanderthalensis* and *Homo sapiens* split.

600,000–400,000 YA: Large carnivores and carcass destroyers go extinct in Europe.

465,000 YA: Age of the oldest hearth yet found in Europe (Brittany).

400,000–350,000 YA: *Homo erectus* or *Homo heidelbergensis* present at Bilzingsleben, eastern Germany.

300,000–250,000 YA: Appearance of *Homo neanderthalensis* in western Europe.

250,000–40,000 YA: Neanderthals adapt to most European environments, during both pleniglacial and interglacial phases, through a variety of hunting techniques and a material culture based on stone tools.

40,000–28,000 YA: Colonization of Europe by *Homo sapiens*. Stabilization of ice in Europe, followed by frequent oscillations of average temperatures. Modern humans adapt with a wide variety of stone tools in addition to artifacts of carved bone, ivory, and antler.

30,000+ YA: Oldest cave paintings in western Europe.

28,000 YA: First open-air settlements in Europe.

20,000–16,000 YA: Last glacial maximum. Mile-high ice renders northern Europe uninhabitable. Many species, including humans, migrate to southwestern Europe.

20,000 YA: Humans tailor fur suits with bone needles in eastern Europe.

15,000 YA: Humans begin to migrate out of southwestern refugia. Beginning of the Mesolithic Era.

10,000 YA: Full retreat of ice. Receding ice leaves northern Europe waterlogged and dotted with lakes in many regions. Loess left over much of north-central Europe.

10,000 YA: Late Pleistocene extinctions of giant deer, woolly rhino, and woolly mammoth in Europe, likely due to overhunting. Mesolithic hunter-gatherers base subsistence on smaller herbivores, fish, and shellfish where available. They actively maintain grassland and forest edge through burning.

8,000 YA: Deciduous forest in place in most of northern and western Europe. Boreal forest in Scandinavia. Full recolonization by mammals, birds, and humans follows the return of forest.

7,800 YA: The North Sea severs Great Britain from continental Europe.

7,000 YA: Maximum temperatures of the Holocene epoch.

5500 BCE: Early Neolithic in north-central Europe. Climate warm and dry. Linear Pottery Culture spreads from the Danube valley.

5000–4000 BCE: Early Neolithic Europeans construct megalithic monuments along the Atlantic fringe.

4400 BCE: Farmers spread out of the loess zone of north-central Europe to the North European Plain and Alpine Foreland.

4000 BCE: Early Neolithic in Scandinavia and Great Britain.

4000 BCE: Elm decline in northwest Europe.

2000 BCE: Stonehenge completed.

1850 BCE: First casting of bronze in central and western Europe. Continental trade in metals accelerates.

1700–1500 BCE: Copper ores exhausted around Salzburg, Austria.

1500 BCE: Banks and ditches used as boundaries around fields in northwest Europe. Agroecosystems assume long-term structure; farmers introduce new grains and crop rotation. Early surpluses of meat and grain exported to Mediterranean regions.

1000–750 BCE: Rapid spread of iron metallurgy across most of Europe. Forests in Britain dwindle by half with respect to Mesolithic forest cover. Animal species dependent on forest also decline.

500 BCE: Rise of Celts in west-central Europe. Horsemanship and iron weaponry become central to Celtic culture.

500 BCE: Iron metallurgy practiced in Scandinavia.

450 BCE: Mass migration of Celts to the south and east. Roman writers attribute the migration to insufficient arable land in the Celtic heartland.

200 BCE: Large enclosed settlements, called "oppida" by Julius Caesar, are constructed from France to Bohemia. They form centers of handicraft production and money-based economies.

58–51 BCE: Roman conquest of Gaul under Caesar.

50 BCE–150 CE: Significant deforestation in southern Germany.

16 CE: Roman Emperor Tiberius begins to withdraw troops to the Rhine, leaving much of northern and central Europe beyond the boundaries of the empire. A trading zone of 200 kilometers extends beyond the Rhine. Northern Europeans feed Roman soldiers and pay taxes to Rome from agricultural surplus. Roads leave deep imprints. Mining and forest clearance produce local effects on northern European ecosystems.

43 CE: Roman Emperor Claudius invades England.

300–700: Demographic decline accompanies the decline of the Roman Empire and subsequent political chaos.

375: Visigoths cross the Danube, accelerating the collapse of the Western empire.

Early 400s: Vandals, Alans, and Suebi overrun the Rhine frontier.

535: A dust-veil event, possibly due to volcanic eruptions, produces five to fifteen years of disrupted weather, failed harvests, and epidemics across much of the northern hemisphere.

700–1200: Populations of south Baltic sturgeon decline due to overfishing.

800–1300: Most western European farmers intensify cereal growing to permanent arable with crop rotations: this is the great age of forest clearance to enlarge the arable surface. In east-central Europe and Scandinavia, permanent woodland clearance for arable peaks in the thirteenth and fourteenth centuries. From the year 1000 on, Europeans gradually rely more on sheep and goats for meat over pigs and cattle, the latter two animals (especially pigs) being dependent on woodland forage.

900–1250: Medieval Warm Period. Slightly higher temperatures than in preceding or succeeding periods allow Vikings to establish colonies in Greenland and farmers to plant vineyards in England.

1085: Officials of King William I of England produce the Domesday survey, a uniquely detailed listing of lordships (fiefs, manors) theoretically everywhere in England, a range of coverage unknown elsewhere in medieval Europe. Particulars let later historians draw important inferences about regionally distinctive density of human settlement and use of a wide variety of land, agricultural resources, and wild resources. The survey shows a forest cover of only 15 percent in mainland Britain.

1200: Owners of high forest begin to protect their timber resources more vigorously.

1200+: Height of medieval urbanization. Growing cities deepen their ecological footprints by importing food and wood from widening hinterlands.

1219: French King Philippe-Auguste creates the *Eaux et Forêts* to manage royal forests.

1300: Woodland in France at 13 million hectares, reduced by more than half since 800 CE.

1300+: Parts of Holland subside below sea level as a result of several centuries of drainage for farming, causing soil compaction. Large windmills pump out water.

1300–1340: Peak in western European populations.

1300–1850: The Little Ice Age brings sharper seasonal differences and mean annual temperatures up to 1.5°C lower than previous Holocene averages.

1315–1319: An unusual sequence of cold and wet weather in northwestern Europe causes repeated crop failures, epizootics, and widespread famine with resulting regional human mortalities approaching 20 percent. These years initiate a long period of environmental and demographic catastrophes during the fourteenth century.

1340s–1370s: Alpine glaciers reach greatest extent during the Little Ice Age.

1341–1362: Norse Greenlanders abandon the Western Settlement under pressure of shorter growing seasons.

1347–1351: The Black Death, a widespread and deadly disease, begins in central Asia or northern China, eventually covering nearly all of Europe. Local mortality rates approach 80 percent and average between one in two or one in three of the population. Other epidemics, at first widespread then regional, recur in the later 1300s and for several centuries thereafter. A sharp reduction in human numbers, combined with higher per capita wealth among survivors, shifts the size and quality of human pressures against land and other resources. The twentieth-century identification of the disease with bubonic plague (and all the ecological connections it might imply) is now doubted by many historical experts, though none so far agree on an alternative diagnosis.

1376: French King Charles V's forest ordinance requires seed-bearing trees to be left following cutting.

Late 1300s+: Commoners are gradually forbidden to hunt game.

1492–1493: Christopher Columbus's first voyage. The major historical exchange of crops, animals, weeds, vermin, and diseases between Europe and the Americas begins.

1500: The Netherlands and Scotland are largely deforested.

1500–1650: Population growth and price revolution in Europe. Strong intensification of agriculture.

1500–1700: Several northern European nations contest rights to fish for cod in Icelandic waters.

1500–1800: European iron production triples.

1550–1789: France's forest cover declines by half, from 18 million to 9 million hectares.

1557: Physician Georg Agricola publishes *De re metallica*, probably the first treatise to expose the environmental hazards associated with mining.

1590–1650: Great wave of land reclamation in the northern Netherlands.

1618–1648: The Thirty Years War in central Europe kills millions of people. Demographic decline is followed by a temporary recovery of wildlife.

1627: The last known auroch (native European wild ox, *Bos primigenus*) is killed as a trophy in its final remnant range in Poland.

1645–1715: Disappearance of sunspots further lowers average mean temperatures in the Northern Hemisphere.

1650–1750: Stagnation and regional decline in northern Europe's population growth. Extensification of agriculture.

1665: Outbreak of the plague in London.

1666: The Great Fire of London destroys the city and ushers in the era of building with brick instead of wood.

1669: A French Forest Ordinance under King Louis XIV attempts to place all forests under royal control, determine methods of cutting, forbid grazing in forests, and curtail common rights to gather wood. The ordinance is enforced with difficulty, and not at all in the Pyrenees.

1696–1697: Extreme weather combined with an absence of public assistance causes one-third of the Finnish population to perish through hunger and disease.

1697: A notorious ordinance under Christian Ernst, Margrave of Brandenburg, seeks to exterminate nearly all forms of wildlife.

1698: Englishman Thomas Savery patents the first coal-burning, steam-powered pump for use in mines. Steam engines diffuse slowly in English mining and industry over the next century.

1700+: Epidemics decrease in number and severity, largely through use of quarantine.

1700s: A total of 445 species of trees and shrubs are introduced in England, mostly to adorn "landscape gardens" of the wealthy.

1713: Carl von Carlowitz writes the first book in German, *Sylvicultura Oeconomica*, concerned exclusively with silviculture and forest management.

1714–1720: Rinderpest kills millions of cattle all over Europe. The epizootic strikes again in the mid-1740s and the early 1770s.

1750–1800: Population growth resumes, accompanied by strong intensification of agriculture.

c.1750: British forest cover reaches a new low of about 800,000 hectares, although the preeminent British navy relies heavily on timber from North America.

c.1750: Approximate 5 percent of Jutland consists of sand dunes, products of deforestation and overgrazing.

1760–1850: The enclosure movement reaches its zenith as the British Parliament enacts the privatization and transformation of one-third of the land in England and Wales into enclosed, rectangular fields more amenable to intensive agriculture.

1760+: Land reclamation in France receives royal backing.

1769: James Watt patents an efficient steam engine capable of producing rotary power, thus adaptable for use in mechanized manufacturing.

1785: First steam engine installed in a cotton mill, located in Nottinghamshire.

1789–1799: French Revolution. The French state appropriates 1.5 million hectares of forest from the Roman Catholic Church and political émigrés. Conquest across the Rhine introduces French foresters to German forestry. Collective rights are protected in the Rural Code of 1791, and hunting is proclaimed a democratic right. The partition of common lands is allowed from 1793 to 1796, but carried out in few places. Communal and private appropriations of land lead to increased clearance for agriculture.

1790s: Canal construction peaks in Great Britain.

Late 1700s: Londoners burn one million tons of coal annually for domestic heating.

1798: Thomas Malthus publishes his famous *Essay on the Principle of Population*.

Early 1800s: Mechanized industry is established in coal-rich areas of western continental Europe.

1800–1850: German peasants are freed of remaining seignorial obligations upon payment of compensation, effectively transferring much land to former lords and allowing planting of large monocultures.

1810: Napoleonic decree classifies French industries according to their "insalubrious" or "disagreeable" odors. Decree establishes allowable distance from human habitation for each category of industry.

1815: The Congress of Vienna creates the international Rhine Commission, whose mandate is to eliminate human and natural barriers to trade along the river. The Rhine will eventually be transformed into a shipping channel with virtually no floodplain.

1815: Eruption of Tambora in Indonesia, causing a dust-veil event in the western hemisphere by 1816, followed by harvest failure, localized famine, and epidemics until 1819.

1817–1876: In the name of flood control, the Tulla Rectification Project eliminates oxbows and loops along the Upper Rhine River, reducing its length by 82 kilometers between Basel and Mannheim and providing it with an artificial bed.

1820–1850: Peak in conversion of coppice to high forest in Europe for purposes of shipbuilding.

1825: English engineer George Stephenson completes the first steam-driven locomotive for the Stockton and Darlington Railway, inaugurating the era of railroad construction. Early railroad construction elicits popular protests against the filling of valleys and cutting through hills.

1827: The French Forest Code submits communal forests to state management, bans grazing of sheep and goats in forests, and provides for efficient repression of "forest crimes."

1830: An Inspectorate of Historical Monuments is established in France to inventory and safeguard monuments and objects of cultural value. The notion of "national heritage" will be extended in the early twentieth century to include certain landscapes.

1831: An early example of nature conservation, Britain's Game Act aims to preserve habitat for species useful for "sport."

1831–1832: The first of four major cholera epidemics in nineteenth-century Europe causes 20,000 deaths in Paris alone.

1840: Justus von Liebig provides the theoretical foundations for agrochemistry in his *Organic Chemistry in Its Applications to Agriculture and Physiology.*

1840–1852: The reclamation of the Netherlands' Haarlemmermeer, an 18,000-hectare artificial lake, deprives the city of Leyden of a means of flushing its canals and puts reclamation workers at risk of catching malaria.

1842: Victor Legrand, director of France's state engineering corps, initiates the "Legrand Star," a national railroad system radiating from Paris and maximizing the use of straight lines and low grades, necessitating deep cuttings and high embankments.

1845: A Prussian Trade Regulation requires potentially harmful industry to obtain licenses from the state following consultation with citizens. Prussian courts place high burden of proof on citizens bringing legal complaints against industry.

1850: The Grammont Law in France punishes public cruelty toward animals.

c. 1850: The demographic transition begins in western Europe.

c. 1850: German forester Carl Obbarius, living in Sweden, cautions against clear-cutting for the sake of biodiversity, leading the way to the decline of clear-cutting in Sweden.

1851: London's Great Exhibition showcases British industry in the Crystal Palace, erected for the occasion in Hyde Park and made exclusively of iron and glass.

1852: An Austrian Forest Act places all communal forests under state supervision and prohibits large cuts.

1852–1870: Baron Haussmann masterminds the renovation of Paris, including new systems to supply the city with fresh water and drain it of wastes.

1853: Britain's Smoke Nuisance Abatement Act is passed under Lord Palmerston.

1858: London's "Great Stink" sets in motion plans for an adequate sewer system.

1859: Charles Darwin publishes *The Origin of Species*, a work that will eventually cause a sea change in both scientific and popular understandings of evolution and the relationships of species to their environments.

1860: France passes an alpine reforestation law, allowing for both voluntary and mandatory replanting, in the name of public utility. Prevention of lowland flooding is the main issue at hand.

1860s: Tall smokestacks become the norm in European industry, quieting local complaints about air pollution.

1862: Louis Pasteur first tests his system of "pasteurization" for killing bacteria in liquids.

1862: The Battle of Hampton Roads in the American Civil War ushers in iron-clad ships, relieving pressure on oak forests for sailing ships. This causes a partial shift in European forestry to short rotations to produce wood for immediate use.

1865: London's sewer system is inaugurated. Sewage is discharged twenty-two kilometers downstream, untreated until the late 1880s.

1865: Urban reformers found the Commons Preservation Society in England to secure and maintain access to open land.

1866: Ernst Haeckel, a German scientist and follower of Darwin, coins the word "Oecologie" to designate the branch of biology specifically concerned with how organisms relate to their "external world."

1869: Paris begins to divert some of its sewage water to Gennevilliers for "sewage irrigation." Small farmers benefit.

1870s+: Officials in Oslo cover malodorous urban rivers, creating a closed sewer system from natural rivers.

1872: The term "acid rain" first appears in print in England.

1873: London smog causes 700 excess deaths.

c. 1875+: Competition from North American grain contributes to agricultural depression in Europe. Surviving farmers are given a greater incentive to specialize in products other than grain.

1876: A Swiss Forest Police Law attempts to control use rights and cuts, while initiating replanting schemes in Alpine and pre-Alpine areas.

1876: A Dutch industrial nuisance law gives municipalities permission to grant licenses for new factories; municipalities are allowed to refuse licenses in case of probable danger to health or property, but industry largely ignores the law.

1879: The City of Manchester wins the right to divert water from Lake Thirlmere.

1880: Forest cover in historical Flanders, extending from Dunkirk to Antwerp in Belgium, is down to a mere 6 percent. Thereafter, grain imports allow reforestation programs.

1880: A Useful Animal Act is passed in the Netherlands to protect animals employed in agriculture and forestry.

1880s+: Mountaineering clubs become popular in western Europe.

1882: Third-Republic legislators in France define "restoration" and "conservation" in the context of a new law to promote alpine reforestation.

1885: The International Conferences on Hygiene finalize procedures for determining the bacteriological content of water.

1886: Gottlieb Daimler and Nikolaus Otto unveil the first four-wheeled vehicle powered by a gasoline engine at their workshop near Stuttgart.

1889: The Society for the Protection of Birds founded in Britain.

1890s: Golf courses become popular in England.

1892: Outbreak of cholera in Paris.

1894: Parisian households forced by law to hook up to the sewer system.

1895: Founding of Britain's National Trust for Places of Historic Interest or Natural Beauty. Parliament makes its property inalienable in 1907.

1896: Hamburg builds Germany's first garbage incineration plant.

1896: Swedish chemist Svante Arrhenius publishes "On the Influence of Carbonic Acid in the Air upon the Temperature of the Ground," the first scientific treatise establishing the theory of anthropogenic global warming.

Early 1900s: A private company in Berlin experiments with waste sorting and recycling but goes bankrupt in 1912.

1904: Germany's Union for Homeland Protection is founded.

1905: Society for the Preservation of Nature is founded in the Netherlands to identify "remarkable" natural monuments and to obtain funds for their purchase and management.

1906: The Public Office for the Conservation of Natural Monuments is established in Prussia.

1906: The French Parliament passes a law on the Protection of Natural Sites and Monuments in order to preserve picturesque landscapes.

1908: The Swedish Association for Nature Conservation is founded.

1909: Sweden designates its first national parks.

1912: The Society for the Promotion of Nature Reserves is founded in Britain.

1913: Prussian politicians decline the opportunity to control the discharge of industrial wastewater into the Rhine. The German chemical industry's position prevails, claiming that a significant portion of the Rhine would have to remain a "sacrificed stretch."

1914: France designates its first, small national park in the Alps.

1914–1918: World War I. Combat on the western front destroys agriculture, towns, industry, and 200,000 hectares of forest in northern and eastern France.

1917: A French industrial nuisance law gives an existing corps of labor inspectors the task of environmental inspection.

1918: The Weimar Republic makes preservation of natural monuments a constitutional guarantee.

1919: The Forestry Commission is created in Britain with a mandate to plant forests for timber.

1923: The ecology of the Ruhr Basin experiences a temporary respite from heavy industrial pollution during strikes following French occupation.

1926: Patrick Abercrombie founds the Council for the Preservation of Rural England.

1927: First official celebration of a national *Fête de l'Arbre* (Arbor Day) in France.

1930s: Conversion of land to urban space peaks in Britain, until the postwar era.

1932: France's Morizet Law prohibits industrial emissions destructive to human health, agriculture, the preservation of monuments, and natural beauty.

1934+: Construction of autobahns (superhighways) in Nazi Germany provides jobs for nearly one million workers and stimulates automobile production. Landscape architects work with highway engineers to assure replanting of trees and conservation of topsoil.

1935: Imperial Nature Protection Law mandates nature reserves and other conservation measures in Nazi Germany.

1939–1945: World War II brings massive destruction to urban, industrial, and agricultural infrastructures. Approximately half of all casualties are civilians. Populations in parts of occupied Europe experience hunger and starvation toward the end of the war. Massive use of natural resources for the war effort.

1947: The Town and Country Planning Act puts British municipalities in charge of granting or withholding permission for changes in land use. Green belts around some cities are one result.

1947–1982: About 200,000 kilometers of hedgerow (42 percent) are cleared in Britain.

1948: The Marshall Plan injects European economies with capital for rebuilding and modernization of industry and agriculture.

1949: The National Parks and Access to the Countryside Act in Britain establishes the framework for creating national parks and managing nature reserves.

Late 1940s–early 1970s: The greatest economic boom in European history begins, anchored in low oil prices. It later becomes known in France as the "thirty glorious years," in Germany as the "economic miracle." The final wave of rural depopulation in modern times gets under way.

1950+: Suburbanization and consumerism deepen the ecological footprint of the European middle class.

1950+: Fishing fleets become equipped with sonar and seine netting, vastly increasing the catch.

1950–1975: Cars become an ordinary means of transportation in western Europe.

1950s–1990s: The number of farms in France decreases by half; three million hectares are subtracted from cultivation.

1951: Swedes Ruben Rausing and Erik Wallenberg found Tetra Pak, a company that will become one of Sweden's largest corporations. In 1963 they introduce the nonrecyclable, rectangular Tetra Brik, a paperboard carton coated with plastic, used for storing milk and other liquids.

1952: London's "Killer Smog" in December causes 4,000 excess deaths and leads to passage of the Clean Air Act in 1956.

1956: Jacques-Yves Cousteau's first feature film about oceans, *The Silent World*, wins the grand prize at the Cannes Film Festival.

1957: The Treaty of Rome establishes the European Economic Community. Its initial members are Belgium, the Netherlands, Luxembourg, Germany, France, and Italy.

1960s+: The ecology of the Alps is increasingly threatened by record numbers of hikers and skiers. Avalanches and floods become more frequent.

1961: The British government enacts a "voluntary" ban on the use of organochlorine pesticides.

1962: The European Commission finalizes its Common Agricultural Policy: direct payments to farmers boost productivity, paving the way for agribusiness, grain surpluses, and increased use of biocides.

1962: Norway designates its first national park, Rondane.

1964: A West German air quality regulation begins to determine threshold values for various air pollutants, defined in terms of human health.

1967: Crude oil from the grounded oil tanker *Torrey Canyon* blackens the coasts of Cornwall, Brittany, and Guernsey, killing some 25,000 seabirds.

1968: A massive student uprising in Paris launches numerous grassroots movements, including contemporary environmentalism in Europe.

1969: Greenpeace and Friends of the Earth are founded. Both organizations will attract substantial followings in much of western Europe.

1969: The Netherlands implements a law to combat water pollution based on the "polluter pays" principle.

c. 1970: Herring nearly become extinct in Scandinavian waters. Populations will recover somewhat by the 1990s.

Early 1970s: Beginning of deindustrialization in parts of Europe, especially textile-producing regions.

1971: West Germany prohibits the use of DDT.

1971: France creates the first Ministry of the Environment in Europe.

1971: First demonstration against a nuclear reactor in France.

1972: Stockholm sponsors the United Nations Conference on the Human Environment.

1972: The Club of Rome publishes *The Limits to Growth*, warning of dire environmental consequences from continuing high rates of economic growth.

1973: Drilling for oil in the North Sea accelerates following the OPEC oil embargo.

1973: The European Community initiates its first Action Programme on the Environment.

1973–1988: Green parties form throughout northern and western Europe.

1974: France launches a massive civilian nuclear energy program.

1974: Agronomist René Dumont runs for the presidency of France on a green agenda.

1974: Salmon are sighted at West Thurrock on the Thames for the first time in a century.

1975–1980: Heyday of antinuclear activism in West Germany. Activists opt for nonviolent tactics by the end of the 1970s and thereby increase their support from the media.

1979: The European Community begins to regulate pesticide use.

1980: The World Conservation Strategy introduces the concept of sustainable development.

1980+: The peace movement helps fuel environmentalism, especially in West Germany.

1982: The International Whaling Commission issues a moratorium on the hunting of all whales, taking effect in 1985–1986.

1986: Chernobyl nuclear accident in the Ukraine. Radiation first detected in Sweden.

1986: An epidemic of bovine spongiform encephalopathy (BSE, or "mad cow disease") is discovered in British cattle, most likely caused by the feeding of animal byproducts to cows. The transmissibility to humans of the fatal neurological disease is discovered in 1996, causing the European Union to ban the export of British beef until 1999. By 2003, 143 of the 152 human deaths due to the disease's human variant have occurred in Britain.

1987: The Brundtland Commission issues *Our Common Future*, a report that will make "sustainable development" a common phrase.

1987: The Norwegian government places sustainable development at the helm of its environmental policies.

1987: The Rhine Action Plan for Ecological Rehabilitation builds aggressively upon earlier international treaties to reduce industrial discharges into the river. Reductions of 50 to 80 percent are achieved for over half of the chemicals targeted. Plans for the restoration of salmon populations and small portions of the Rhine's former floodplain follow.

1989: Revolutions in central and eastern Europe begin to reveal environmental legacies of state communism.

1989: Approximately three-quarters of the world's 400 million tourists choose a European destination.

1990–1991: The *Fundis* leave the German Green Party, allowing *Realos* to move ahead with a reformist agenda.

1991: Following a German government regulation concerning packaging, the company Dual System Germany licenses its symbol, a green point, to manufacturers of recyclable packaging. Consumers are encouraged to purchase items marked with the green point and to dispose of packaging at specific sites. Similar legislation follows in Austria, France, Sweden, and the European Union.

1992: United Nations Conference on Environment and Development held in Rio de Janeiro.

1992: The Maastricht Treaty creates the European Union. It consists of the six founding members of the EEC and Great Britain, Ireland, Denmark, Greece, Spain, and Portugal.

1992: The European Union forbids its members to catch fish in the Barents Sea in hopes of restoring the cod fishery.

1993: Norway resumes a commercial catch of minke whales in the northeast Atlantic.

1995: Sweden, Finland, and Austria are admitted to the European Union.

1995: One out of every ten Britons belongs to an environmental organization.

1997: The Treaty of Amsterdam places sustainable development at the center of all EU policymaking.

1997: Greens and Socialists form a governing coalition in France.

1998: Greens and Social Democrats form a governing coalition in Germany.

1999: French sheep farmer and peasant activist José Bové leads a peaceful dismantling of a McDonald's restaurant under construction in Millau, France, catapulting his organization, the Confédération Paysanne, to the center of a movement calling for small-scale, sustainable agriculture.

1999: Hurricanes Lothar and Martin devastate forests in broad areas of southwestern and northern France. Historians of climate call them the worst hurricanes to hit France since the 1730s.

2000: The German government decides to end its civilian nuclear energy program.

2001: Nearly one out of every four Dutch people belongs to an environmental organization.

2002: The European Commission adopts a more stringent program of regulating pesticides, with provisions for rewarding pesticide-free farming.

2003: The European Union reforms the CAP. In the future, funds will be increasingly directed toward less intensive, more environmentally friendly agriculture.

2004: The journal *Science* documents drastic declines in the populations of birds, butterflies, and native plants in Britain over the last several decades.

2004: The European Union acquires ten new members: Cyprus, Czech Republic, Estonia, Hungary, Latvia, Lithuania, Malta, Poland, Slovakia, and Slovenia. New members have generally weaker environmental records than older members.

2004: Delegations from several European Green parties found the "European Greens."

2004: The last functioning coal mine in France closes.

GLOSSARY

ACID RAIN. The deposition of acids via precipitation, often far from their source, produces rain with a low pH value. Sulfur dioxide and, secondarily, nitrogen oxides contribute most to acid rain; they originate from the burning of fossil fuels and biomass, and also from metallurgy. The expression was first used in a report on atmospheric pollution titled *Acid and Rain,* published by Englishman Robert Smith in 1872. A transnational conflict over acid rain occurred during the 1970s and 1980s between the United Kingdom and Scandinavian countries. At the United Nations Conference on the Human Environment, held in Stockholm in 1972, Swedish scientists presented findings that British industrial emissions were acidifying Scandinavian lakes and forests. When British scientists in turn produced studies of acid rain, their conclusions did not disprove those of Swedish scientists but showed that acidification was also occurring in areas of high rainfall within Britain itself. Acid rain achieved even wider public recognition in the 1980s, in particular when the German media associated acid rain with *Waldsterben,* or "forest death." However, more recent studies have refuted the hypothesis of long-term forest decline in Germany. Many causal factors contribute to the degradation of forests and lakes in northern Europe, but it remains clear that acid deposition alters the chemistry of bodies of fresh water and compromises their biodiversity.

ACTION PROGRAMME ON THE ENVIRONMENT. The framework in which the European Union makes environmental policy. The first Programme was initiated by the European Community in 1973 and focused on pollution; just four years later, however, the second Programme introduced broader principles, such as international collaboration and the "polluter pays" idea, and new mechanisms such as the Environmental Impact Assessment. The European Environmental Agency oversees the implementation of environmental directives in the member states, and a European Environmental Bureau acts as a liaison between voluntary associations and the European Union.

AGENDA 21. This political program was the formal outcome of the United Nations Conference on Environment and Development held in Rio de Janeiro in

1992. Described in a lengthy document, Agenda 21 included the "Rio Declaration," which outlined, in twenty-seven principles, a program of international cooperation for the purpose of sustainable development. Agenda 21 called upon states to eradicate unsustainable forms of production yet explicitly placed greater responsibility upon develeoped countries, whose environmental standards might be "inappropriate and of unwarranted economic and social cost to other countries, in particular developing countries" (Principle 11). The document also required states to apply the *precautionary principle* "according to their capabilities" (Principle 15), and it linked principles of sustainable development to the promotion of global peace. More than 178 countries agreed to help promote Agenda 21.

AGER, SALTUS, SILVA. These Latin terms were used by Roman writers to designate the three classic elements of European agroecosystems, which assumed their long-lasting historical structure during the Bronze Age. *Ager* refers to cultivated fields, whether farmed on a permanent or temporary basis. Fields could be bounded by banks and ditches, as in Britain from 1500 BCE. At that time the introduction of new crops, manuring, and possibly crop rotation complexified and helped sustain the productivity of the *ager*. The *saltus* was pastureland, sometimes bounded by ditches as well. The size of pastures relative to cultivated fields remains unclear for the prehistoric centuries; many of the enclosed "Celtic fields" of Britain may have been pasture. The *silva* was woodland, vital to the agricultural economy by virtue of its many resources—wood, fodder, wild plants, grazing land for larger animals, and habitat for pigs. *Ager, saltus,* and *silva* were articulated in a system of domestic subsistence production, although agricultural production became increasingly oriented toward surplus production for commercial exchange under the influence of the Roman Empire.

APPROPRIATE TECHNOLOGY. The OPEC oil embargo of 1973 caused a rude awakening in the developed world as to its dependence on imported fossil fuels. As a result, "appropriate technology" first came to describe alternative ways of producing energy, such as harnessing wind and solar power. As the environmental movement developed during the 1970s, the meaning of the phrase extended to all forms of technology deemed sustainable and appropriate for given economies, skill levels, and resource bases. For example, ease of use and repair are key criteria for appropriate technology. Some technologies that reduce reliance on energy or paper, however, qualify as "high technology," and thus "appropriate technology" remains difficult to define, much less obtain. References to it often occur in the context of technology design for developing countries.

ARRHENIUS, SVANTE AUGUST (1859–1927). A Swedish chemist best known in his time for his theory of electrolytic dissociation (for which he won the Nobel prize in 1903), Arrhenius is considered to be the scientific "founding father" of global warming. In 1896 he published an article on the effects of atmospheric carbon dioxide on earth's surface temperatures and went on to formulate a model by which to calculate temperature changes based on given variations in the concentration of carbon dioxide. Though aware of the increases in that concentration due to fossil fuel combustion, Arrhenius concluded that atmospheric carbon dioxide would not double for another three millennia. His Scandinavian perspective seems to have influenced the optimism with which he viewed a gradually warming climate.

BOVÉ, JOSÉ. Militant cofounder of and international spokesperson for France's Peasant Confederation, Bové is best known for his acts of civil disobedience against global agribusiness and fast food. His career has included raising sheep on France's remote Larzac plateau, where he helped block the extension of a military camp during much of the 1970s; antinuclear activism; and antiglobalization activism in the context of recent meetings of the World Trade Organization. Bové proposes an alliance of small farmers, consumers, and environmentalists in order to further sustainable agriculture that would assure a livelihood for small producers around the world and a diverse, high-quality food supply.

BRUNDTLAND COMMISSION. Formally the World Commission on Environment and Development, it was appointed by the United Nations under the chairmanship of Norwegian Prime Minister Gro Harlem Brundtland in 1983. Four years later its highly influential report, titled *Our Common Future*, cataloged environmental crises, described them as interlocking, and linked them to global poverty and international insecurity. In calling for a new era of international cooperation, the report defined the concept of sustainable development. It also deepened public understanding of the connections between economic development and the environment and helped frame discussions that led to the United Nations Conference on Environment and Development, the "Rio Conference" of 1992. See *sustainable development*.

CELTS. A people so named by Greek writers, whose identity emerged in the early fifth century BCE. The geographical center of Celtic culture was the Marne-Moselle region of west-central Europe, where early contacts with the Etruscans produced a new warrior elite that ultimately supplanted the Hallstatt culture based in Alpine Europe to the south. The Celts produced a distinctive style of art, "La Tène," named after a site at Lake Neuchâtel, Switzerland, with curvilinear forms loosely influenced by Greek and Etruscan art. The Celtic language spread to the Atlantic seaboard and along the rivers flowing into the Atlantic. An Iron Age people, the Celts developed sophisticated agroecosystems marked by the use of iron implements. A significant Celtic migration to the south and east occurred in the fourth and third centuries BCE; though the image of raiding parties seeking plunder was long-lasting in Roman culture (Celts sacked Rome itself in 390 BCE), most migrants left in search of new land to farm, possibly indicating the insufficiency of agriculture relative to population in the Celtic heartland. Archaeological evidence does not, however, provide clear proof of massive and permanent out-migration.

CLIMATE CHANGE. Alterations in earth's climate, which may amount to changes in temperature, precipitation, cloud cover, and/or other aspects of climate—usually with distinctive regional manifestations—have been attributed to natural factors until the recent phenomenon of anthropogenic global warming. Variations in solar activity, the earth's orbit, and volcanic activity can all trigger climate change by altering the amount of solar radiation that reaches earth. The two instances of climate change to have affected northern Europe since postclassical times have been the Medieval Warm Period (c. 900–c. 1250 CE) and the Little Ice Age (c. 1300–c. 1850). Most of the global warming observed since about 1860 probably stems from the accumulation of carbon dioxide and other gases in the atmosphere, largely a result of fossil fuel combustion and deforestation. See *Arrhenius, Svante August.*

CLUB OF ROME. This think tank was founded in Rome in 1968 by an international group of scientists, economists, businessmen, and civil servants. It sought, and continues to seek, innovative ways in which to address various issues of global import. The Club's most famous publication, authored by Dennis L. Meadows, Donella H. Meadows, and Jorgen Randers, was *The Limits to Growth* (1972). The authors, trained in computer modeling and systems dynamics, explored a variety of variables in their analysis of global environmental decline. Analyzing the interdependencies and *positive feedback* among industrialization, population growth, malnutrition, rapid exploitation of nonrenewable resources, and degraded ecosystems, they concluded that present growth trends could, if unaltered, continue for at most another hundred years before triggering population decline and industrial collapse. Criticized for some of its assumptions regarding the availability of resources and the global nature of the model, *The Limits to Growth* has, on the other hand, been supported by other research. The authors published a subsequent study titled *Beyond the Limits* in 1991. Their conclusions were markedly similar to those in the 1972 report, but here they focused more on the transition to sustainable society. Across the political spectrum, however, economic growth remains a fundamental goal in western democracies.

COEVOLUTION. An evolving and mutually influential process of interaction between components of an ecosystem, such as between a human cultural population and its natural environment. Coevolution has been observed, for example, between pollinators and the plants they pollinate, and between human cultures and their domestic plants and animals.

COMMON ACCESS. Common access pertained to communal property held for purposes of grazing, haying, and the gathering of firewood and, more generally, to collective use rights to land or other resources (water, etc.) not necessarily under collective ownership. Typical use rights included grazing, gathering firewood, gleaning, and pasturing animals on harvested fields and fallow land. Depending on location, use rights could obtain on land belonging to another village, land owned by an individual, or land belonging to the Crown or the state. They tended to be strictly regulated by communal authorities, although abuse could follow from demographic or social instability. Usurpations of common land were relatively common in the early modern era. The transition to modern agrarian capitalism brought about legal attacks on, and a drastic curtailment of, use rights. However, during the French Revolution the Rural Code of 1791 upheld use rights. A Revolutionary decree of 1793 allowed the partition of communal property at the request of one-third of the inhabitants, but little partition occurred in practice, and the decree was rescinded in 1796. Gradual legal changes in the nineteenth century whittled away use rights in France, as also happened in German-speaking lands following the emancipations of 1848. In England the long process of *enclosure* extinguished common property and access earlier than in continental Europe.

CORE. This term used by world-systems theorists refers to those societies possessing the economic, technological, and military resources to enforce their economic dominance. As centers of transshipment, banking, and sophisticated commodity production, core areas extract resources from peripheral areas (see *periphery*) and, in the case of political empires, exact revenues as well. Most commonly associated with modern world capitalism, the term can also be applied more loosely to ancient trade empires.

COUSTEAU, JACQUES-YVES (1910–1997). Having invented the aqualung in the 1930s, a device that allowed him to move freely at ocean depths, Cousteau achieved fame from the 1950s with numerous feature films about the world's oceans. Through television and film Cousteau introduced a large international audience to underwater life. His films combined adventure with science and bore witness to environmental degradation caused by pollution and overfishing. Cousteau kept his distance from French environmentalists, refusing their invitation to run for president as a Green in 1981; he also appeared to sanction the French government's nuclear tests on Mururoa Island in the Pacific in 1987.

DEEP ECOLOGY. Norwegian philosopher Arne Naess coined this term in 1973 in order to redirect the environmental movement away from its anthropocentric assumptions. In subsequent years, "deep ecologists" identified themselves through their concern with human overpopulation and their insistence that *Homo sapiens* be submitted to the same methods of analysis that the science of ecology applied to all other species. For many deep ecologists, a biocentric perspective had a spiritual dimension. They influenced Green parties and mainstream environmentalism relatively little, though they did arguably succeed in diversifying the movement as a whole.

DEMOGRAPHIC TRANSITION. The decline of mortality, due to improvements in diet and sanitation, followed by declining fertility and a drop in family size to two to three children. In societies undergoing the transition, birth and death rates drop from thirty to thirty-five per thousand per year to ten to twelve per thousand per year. Broadly associated with changes wrought by industrialization, the demographic transition began in western Europe in the mid-nineteenth century but took over one hundred years to complete itself. As a model, the demographic transition has shown itself to be only partially applicable to twentieth-century societies undergoing economic development.

DESERTIFICATION. A severe instance of land degradation resulting in the formation of deserts due to human activities, usually excessive tilling of soil or overgrazing, which leads to soil compaction and erosion. Sometimes climate change, namely prolonged drought, accelerates desertification. Arid and semi-arid ecosystems are most vulnerable to desertification. Northern Europe's temperate climate and (over the long term) fairly even precipitation have largely protected this region from desertification, although deforestation on some mid-altitude slopes in alpine regions has led to substantial erosion. These bare slopes did evoke an image of desertification for many foresters in the modern era.

DOMESTICATION. The gradual process, beginning around 11,000 BCE with dogs, by which human groups came to control certain species of animals and plants through selective reproduction. Plant domestication, first involving annual varieties of pulses and grains, happened initially in the Middle East and provided the most basic condition for agriculture. The *LBK* culture first brought domesticated plants to northern Europe. Domesticated goats, sheep, pigs, and cows also came to Europe from the Middle East and western Asia.

DOWER, JOHN (D. 1947). An architect by training, Dower became a leader in the British movement to create national parks prior to and during World War II. His report titled *National Parks in England and Wales*, published in 1945, served to inform the National Parks Commission, which created ten parks during the 1950s. Dower believed that the promotion of public access to the parks could be a tool in their conservation, and that farming must be respected as part of a broader conservation policy relevant to Britain's "functioning" countryside.

DUMONT, RENÉ (1904–2001). World-renowned agronomist who became the most heralded spokesperson for the French environmental movement in the 1970s. His candidacy for president in 1974 marked the beginning of the movement's politicization. Dumont insisted on the global dimensions of environmental problems and their fundamental nature, writing in 1977, "It is no longer a question of being satisfied with protecting parks and country homes and little birds; we must reinvent our entire civilization." (Whiteside 2002, 31)

ECOLOGICAL FOOTPRINT. The overall ecological impact of a given phenomenon or activity. Used in the context of urbanization, ecological footprints refer to the many ways in which cities affect the ecologies of their hinterlands, from the importation of water, food, and building materials, to the discharge of sewage, garbage, and other effluents. The ecological footprints of many modern cities can be said to crisscross the planet as cities and their suburbs consume resources from distant sources and contaminate extensive portions of water, air, and soil with exported wastes.

ECOLOGY. Northern European scientists have considerably influenced the development of this science, both long before and after the German Ernst Haeckel coined the term in 1866. The work of several generations of naturalists, geographers, and biologists helped spawn ecology. Gilbert White, Carolus Linnaeus, Jean-Baptiste Lamarck, and Charles Darwin, among other northern Europeans, must be included among ecology's forerunners. Classification and geographical plotting of species were key endeavors that helped ground ecology's chief concern—the interrelationships in nature. The Danish plant geographer Eugenius

Warming helped place ecology on a scientific footing by exploring what he called "communities" of organisms, specifically forms of mutualism including commensalism and symbiosis, and the succession of communities toward "climax" formations—an idea that would become highly influential on both sides of the Atlantic. In Switzerland and France, scholars of the Zurich-Montpellier school worked to classify plant communities. In 1935, British botanist Arthur Tansley coined the term "ecosystem," a concept that became ecology's primary unit of analysis. For Tansley, substituting "ecosystem" for "community" represented a necessary turn away from the anthropocentric implications of the latter term: ecosystems were essentially sites for the circulation of energy and chemical substances, and their study should be hitched to the more established sciences of physics and chemistry. In the same era, British zoologist Charles Elton pioneered in animal ecology through his analysis of "food chains" and "niches." The advance of ecology in the early twentieth century can be charted also by the formation of such institutions as the British Ecological Society, founded in 1913.

EEDEN, FREDERIK WILLEM VAN (1829–1901). A Dutch biologist and civil servant, van Eeden was an early conservationist and probably the first Dutchman to advocate the preservation of "monuments of nature." His advocacy came as a reaction to the cutting of the Netherlands' last pristine forest in 1871. His brand of conservationism was in line with that of many German, French, and Scandinavian conservationists of the late nineteenth century who sought to preserve small, often isolated remnants of preindustrial landscapes for reasons of national heritage, as opposed to preserving viable ecosystems. Van Eeden also played a role in animal protection; from 1870 he called for the protection of animals deemed useful in agriculture, a demand which came to fruition in the Useful Animal Act of 1880.

ENCLOSURES. In Great Britain, the long transition from strip farming on open fields to grazing and intensive farming on fields enclosed by walls, fences, or hedges. Common lands and "waste" lands were also enclosed. Enclosures took place piecemeal from the twelfth century, accelerated with sheep farming in the fifteenth and sixteenth centuries, and completed the transformation of rural Britain during the century following 1750, as parliamentary acts permitted several thousand new enclosures. During this final wave, close to seven million acres, or one-third of the total land area of England, were enclosed. In addition to the loss of land and use rights by small farmers, enclosures reconfigured the landscape and permitted more intensive exploitation, including mechanization, of farmland.

EQUILIBRIUM. In the natural sciences, equilibrium refers to a state of dynamic balance, not an absence of change but rather a self-correcting system, able to absorb changes arising both within and from outside. Within some limits, a system in equilibrium is self-sustaining. Famous examples of ecological equilibria are the oscillating numerical relationship between predator and prey species, or the self-perpetuating quality of woodlands dominated by shade-tolerant trees like maple, linden, or beech. Ecologists no longer think all systems reach equilibrium, but many do tend toward that state until external pressures become too great, then they change quite radically. The relationship between human populations and their food supplies famously postulated by English economist Thomas Malthus is also an equilibrium model. Many quasi-equilibria result from long *coevolution* among elements of the system.

EUROPEAN ECONOMIC COMMUNITY. The original name of the international organization founded by the Treaty of Rome in 1957. Its six founding members—Belgium, the Netherlands, Luxembourg, Germany, France, and Italy—established a customs union among themselves that facilitated the free movement of goods, services, capital, and people. The EEC was later renamed the European Community and known informally as the Common Market. The organization formally began to make environmental policy from 1973 with its *Action Programmes.* Today the EC remains one of the three "pillars" of the *European Union.*

EUROPEAN UNION. Created by the Maastricht Treaty of 1992 and the result of negotiations among members of the European Community concerning monetary and political union. The European Union launched a common European currency, the Euro, and was enlarged to fifteen members in 1995 and twenty-five members in 2004. The EU's three pillars consist of the European Community (see above), the Common Foreign and Security Policy, and Justice and Home Affairs. Article 130r (2) of the Maastricht Treaty stipulates that environmental poliy must be based on the *precautionary principle* and preventive action and that environmental policy must inform all other policies of the European Community. Under the Treaty of Nice (2000), however, environmental policy remained subject to unanimous decision-making. It remains to be seen whether the new member states with weaker environmental records will choose to block further environmental directives.

EUTROPHICATION. The overloading of aquatic ecosystems with organic pollutants, namely compounds containing phosphorous or nitrogen. These nutrients cause populations of algae and blue-green bacteria to explode and deplete the dissolved oxygen in water, thus destroying fish populations. In Europe eutrophication began to occur with the production of synthetic fertilizers in the nineteenth century. Because agricultural runoff comes from decentralized sources, it is difficult to control and has lowered the biodiversity of such major rivers as the Rhine.

FOREST TRANSITION. The transition from net deforestation to net reforestation, often bringing a substantial increase in forest cover. This transition characterizes most of western and northern Europe of the past 150 years. The forested area of Switzerland expanded from 15 percent to 30 percent of the national territory during this time, for example. Factors in the transition include decreasing rural population, the substitution of alternative fuels and building materials, and the professionalization of forestry.

FRENCH DESERT. Introducing the term in the title of his 1947 book, geographer Jean-François Gravier designated all of provincial France as a "desert," deprived of vitality by the weight of Paris in all cultural, economic, and political domains. Used today, the term more often applies specifically to rural areas marked by demographic decline and land abandonment, products of France's drive toward agricultural modernization since the 1950s.

FUNNEL BEAKER CULTURE. This archaeological term refers to funnel-necked beakers that are the characteristic pottery of the first farming culture in most of southern Scandinavia and the North European Plain, as well as later farming cultures (following the *LBK*) in parts of central Europe. Continuity between the pottery of the Ertbølle, a late Mesolithic culture of southern Scandinavia, and that of the Funnel Beaker Culture may indicate the adoption of agriculture by indigenous hunter-gatherers in the region.

GHOST ACREAGE. Imports of foodstuffs, fibers, and other raw materials that effectively add to the available productive acreage in a given area or country. Early modern and modern Europe disposed of considerable ghost acreage through commercial dominance over and imperial relations with much of the world. For example, Belgian industrialization was abetted by the copper, zinc, and cobalt that Belgian concessionaires extracted from the Congo. Accelerated deforestation and heightened mineral extraction worldwide stemmed from Europe's possession of ghost acreage. See chapter 4 for an extended discussion of this concept in the context of the early modern era.

GODWIN, GEORGE (1813–1888). A nineteenth-century British architect, art critic, and longtime editor of the architectural journal the *Builder*, Godwin lobbied for sanitary reforms and urban renewal. Unlike most analysts of urban social ills in the Victorian age, he attributed many problems, including poor public health, rampant criminality, and overcrowding, to environmental factors. In so doing, he was one of the first observers of the modern city to begin to understand slum conditions and their consequences. His contemporaries were more apt to attribute the human degradation of English slums (particularly the slums of London) to personal failure, sin, or Irish immigration. In addition to championing improved housing for the poor, Godwin believed that small, local parks should be intrinsic to slum renewal. Like *Octavia Hill*, he also advocated access to uncultivated nature for urban residents.

GREAT STINK. In the summer of 1858, drought and high temperatures combined to reduce the water level of the Thames, a river whose concentration of raw human sewage rose to 20 percent that summer. The "Great Stink" caused illnesses in the riverside neighborhoods of London and the deaths of thousands of fish. Parliament's attention ultimately resulted in the acceptance of plans for a system of intercepting sewers, whose chief architect was Sir Joseph Bazalgette. The sewer system, which diverted the discharge of London's sewage fourteen miles downstream, contributed to the extinction of cholera in London.

GREEN REVOLUTION. A set of changes that boosted the efficiency and hugely increased the productivity of agriculture in many parts of the world, though particularly in South and East Asia and Latin America, from the 1960s on. New hybrids of rice, maize, and wheat were typically planted in monocultures that required mechanization and heavy doses of pesticides and herbicides. In evaluating the Green Revolution, gains in productivity must be measured alongside the lack of social equality or food independence that have resulted in the developing world, in addition to declining crop diversity and toxic agricultural runoff.

GREENLAND ICE CORES. These deep ice cores drilled from the Greenland summit have been used since the late 1980s to provide detailed information about past climates from about 100,000 years ago.

HABER, FRITZ (1868–1934). A German-Jewish academic chemist, Haber contributed hugely to agrochemistry by inventing a method to extract nitrogen from air through a process of ammonia synthesis. The artificial synthesis of nitrogen allowed the mass production of nitrates for fertilizer and, later, explosives. The significance of the Haber-Bosch process (Karl Bosch was an industrial chemist) stemmed from the difficulty of obtaining nitrogenous fertilizers in large quantities other than through using guano, a major Chilean import to Germany and other European countries in the nineteenth century. Artificial fertilizers became heavily used in twentieth-century Europe and ultimately in much of the world. They have significantly altered soil chemistry, made agriculture more dependent on fossil fuels, and polluted waterways by causing *eutrophication*.

HILL, OCTAVIA (1838–1912). Most notably a housing reformer in Victorian Britain who worked to acquire and renovate dilapidated buildings inhabited by the poor people of London. Along with a handful of like-minded reformers, she believed that the urban environment could be made more amenable to people's needs, and that the presence of nature inside and outside the city was vital for health and well-being. To that end she helped found the Commons Preservation Society in 1865, which helped secure public access to the unenclosed green spaces around London; the Kyrle Society in 1876, whose mission was to bring slum residents into direct contact with nature; and the National Trust for Places of Historic Interest or Natural Beauty in 1895.

HOLOCENE. The present geologic epoch, dating from 10,000 years ago and associated with an *interglacial*.

"HOMO HEIDELBERGENSIS." A proposed species name for the earliest Europeans. Their remains have been dated to approximately 400,000 years ago.

INTERGLACIAL. Since the onset of glaciations approximately 2.3 million years ago, periods of warmer climate have separated glacial phases. During interglacials temperatures approximated those of today. The present interglacial has lasted about 10,000 years; the previous, or Eemian interglacial, lasted from 126,000 to 118,000 years ago.

INTERNATIONAL UNION FOR THE CONSERVATION OF NATURE AND NATURAL RESOURCES (IUCN). Conceptually dating back to 1913, the IUCN grew out of the International Union for the Protection of Nature in 1956. It presently has over 1,000 members from 140 countries, consisting of governments, government agencies, and nongovernmental organizations. The IUCN functions as a network through which concrete programs for sustainable development receive support. It has been instrumental in gathering and publicizing data on endangered species through its *Red Data Books*.

JUNGK, ROBERT (1913–1994). German-Austrian journalist, author, and "futurologist." His critique of nuclear power and large-scale technology made him an important theorist of the antinuclear, environmental, and peace movements in West Germany and Austria during the 1970s and 1980s.

KOCH, ROBERT (1843–1910). Koch was a German physician and bacteriologist who, along with Louis Pasteur, discredited the *miasma* theory of illness by discovering the microorganisms that cause several diseases, namely anthrax and tuberculosis. He thus inaugurated the "germ theory" of infectious disease and became a Nobel laureate for his work. Koch revolutionized medical science and Western perceptions of the origins of disease. In place of the quasi-mysterious miasmas that emanated from decaying organic matter, physicians and scientists now sought to isolate specific microscopic agents. In a sense, Koch enlarged the Western notion of the environment by revealing the invisible world of microorganisms that can weaken or destroy human life.

LAMARCK, JEAN-BAPTISTE (1744–1829). French naturalist who was one of the first exponents of biological evolution. His ideas concerning inheritance, though they substantially influenced nineteenth-century scientific thought, are largely discredited today. Theorizing that species could become modified in response to habitat, and that acquired characteristics could be inherited, Lamarck shaped both naturalism and ecology with his emphasis on the importance of natural milieu. Charles Darwin's theory of natural selection eventually supplanted Lamarck's ideas, yet rare instances of the inheritance of acquired characteristics have been observed; this phenomenon is known as epigenetic inheritance.

LANCASHIRE SYSTEM. The Pennine Mountains, which extend from the Peak District in the English Midlands north to the Scottish border, show many tortuous folds in which coal seams once lay close to the surface. From the early eighteenth century, mining engineers took advantage of geology to cut costs: their "system" called for sinking multitudes of small, shallow coal pits, thereby avoiding great depth or extensive underground tunnels. The pockmarked landscape that resulted could be found in many Lancashire parishes.

LBK. The German acronym for *Linearbandkeramik,* or Linear Pottery Culture, referring to the distinctive incised lines decorating the nearly spherical bowls of the first farming peoples in central and eastern Europe. Following the spread of agriculture from the eastern to the western Mediterranean, the LBK first appeared around 5500 BCE along the middle Danube. This Neolithic culture spread extremely rapidly, as farming peoples settled in areas of loess from the Ukraine to Belgium. Although Mesolithic acculturation has been advanced as an explanation for the LBK, strong arguments in favor of colonization continue to be made.

LIEBIG, JUSTUS VON (1803–1873). Through his work as an organic chemist, Liebig focused his research on plant nutrients, paving the way for the synthesis of fertilizers (see *Haber, Fritz*). His work *Organic Chemistry in Its Applications to Agriculture and Physiology*, published in both German and English in 1840, helped anchor the emerging field of agrochemistry. One of Liebig's key discoveries was that plants require nutrients from both air (fixed by microbes) and soil. His "Law of the Minimum" states that plant growth is limited by the scarcest necessary resource; though applicable to economies of fertilizer use, this axiom became less relevant in the twentieth century as industrial agriculture supplied (or oversupplied) necessary plant nutrients.

MAGDALENIAN. The final of four cultural periods of the Upper Paleolithic in western Europe. Each period is marked by specific tool types and manufacturing processes. In chronological order they are the Aurignacian (34,000–30,000 years ago), the Gravettian (30,000–22,000 years ago), the Solutrean (22,000–18,000 years ago), and the Magdalenian (18,000–11,000 years ago). Magdalenian artists produced most of the celebrated cave paintings in western Europe.

MECHANIZATION. In its historical sense, the term refers to the substitution of externally powered mechanical devices for human skills. Mechanization was one hallmark of both the Industrial Revolution, which ushered in steam-driven looms, lathes, and locomotives, in addition to other machines, and modern industrialized agriculture, symbolized by the tractor in the twentieth century. Mechanization increased productivity and thereby drove up the environmental costs of both agriculture and manufacturing.

MEGALITHIC MONUMENTS. A variety of structures, ranging from mounds and passage graves to stone circles and rows, constructed during the fifth and fourth millennia BCE. They appear along the diverse landscapes of the Atlantic coastline from southern Scandinavia to the Straits of Gibraltar. They coincide with the Neolithic way of life, yet their precise relationship to Neolithic ideologies or worldviews remains unclear. Recent research attempts to interpret the significance of the ways in which these monuments are situated in, and reflect, their natural settings, in addition to exploring what the monuments may indicate about the rituals and astronomical knowledge of early Neolithic Europeans.

MESOLITHIC. Beginning 10,000 years ago, this was a transitional period between the Paleolithic and Neolithic eras, marked in Europe by the recolonization of the northern parts of the continent toward the end of the last glacial era.

MIASMA. Central to prevailing medical through during most of the nineteenth century, the notion of miasma referred to corrupted air thought to carry pathogens. According to miasma theory, impurities could enter the air via putrefying organic matter. Variants on the theory held that pathogens could be destroyed through deodorization, dilution, or steam heat. Though inexact in its correlations between pollution and disease and ultimately discarded in favor of *Robert Koch*'s germ theory, miasma theory influenced the construction of metropolitan sewer systems that did, in effect, help eliminate epidemics of cholera and typhoid in European cities.

MONOCULTURE. A major trend of modern agriculture defined by the cultivation of a single species in a field at any given time. The growth of monoculture went hand in hand with agricultural mechanization. In Europe, some monocultures became permanent and came to dominate entire regions. Although efficient, monoculture leaves cultigens susceptible to insects and disease and thus requires much chemical assistance by way of pesticides and herbicides.

MOUSTERIAN. Material culture associated with *Homo neanderthalensis* from 250,000 to 50,000 years ago.

NEOLITHIC PACKAGE. A concept developed by archaeologists in order to distinguish Neolithic from Paleolithic assemblages. The presence of pottery and polished stone tools set these assemblages off from earlier ones, but the notion of a full "package" came to include evidence of permanent settlement and, most importantly, remains of domesticated plants and animals in the form of pollen and bones—at a minimum, wheat, barley, cattle, sheep, and goats. Specialists have questioned the concept, as many Neolithic sites have not yielded all components of the "package." In specific regions of Europe, the presence of only some Neolithic elements has fueled the hypothesis that the Neolithic way of life was not introduced in full by migrants, but rather adopted gradually and deliberately by local, Mesolithic populations.

OPPIDA. Julius Caesar coined this term to describe the enclosed, fortified settlements established from the middle of the second century BCE and located from central France to the Sudeten Mountains. The largest built structures of pre-Roman Europe, they provide evidence for the technological, economic, and social complexity of late Iron Age societies. These urban areas were centers of manufacturing and trade with both their immediate hinterlands and other oppida, and they functioned on the basis of a money economy. They became key nodes in the trade networks gradually dominated by the Romans.

ORGANOCHLORINES. The category of chlorine-containing hydrocarbon pesticides, such as DDT, aldrin, and dieldrin, among many others. The dangers of indiscriminately used organochlorines came to light in a seminal work by American scientist Rachel Carson, *Silent Spring* (1962). The wide dissemination and translation of Carson's book galvanized much European discussion and research into the ecological consequences of the use of pesticides.

PALEOLITHIC. The earliest and longest period of human history corresponding to the latter part of the Pleistocene geologic epoch. The Paleolithic began 2.5 million years ago in Africa and 900,000 years ago in Europe. It is conventionally divided into Lower, Middle, and Upper subperiods. The latter corresponds with the arrival of modern humans in Europe.

PASTEUR, LOUIS (1822–1895). Pasteur began his scientific career as a chemist and went on to pioneer microbiology along with *Robert Koch*. One of his most famous demonstrations proved that fermentation resulted from invasions of microorganisms, not spontaneous generation. He was thus able to give a scientific grounding to the concept of "purity," which would have far-reaching implications for the treatment of human food supplies and the purification of water. The process of pasteurization (which he first tested in 1862) requires the heating of liquids in order to kill bacteria and molds. Pasteur also contributed to immunology by creating weakened forms of disease to use in "vaccines," a word he coined.

PASTORALISM. European peasants invented a variety of ways in which to raise livestock from the Neolithic up to the modern era. Seasonal constraints continually forced the question of winter feed on populations that attempted to keep livestock year-round. Between the extremes of slaughtering animals in the autumn and keeping them in winter barns to feed on fodder crops—the modern, "sedentary" mode—the survival of livestock during winter often depended on seasonal migration. Technically, pastoralism involved migration between summer and winter pastures of cattle or sheep within distinct regions. It was frequently practiced in mountainous regions of Europe, where low- and high-altitude pastures lay in relative proximity to each other. A multitude of ways in which to organize pastoral grazing could be found across Europe: family members or professional shepherds would migrate with the animals; pastures would be owned, rented, or accessed through use rights by communities or families.

Because of the shorter distances involved, pastoralism differs from *transhumance*, although some specialists refer to all seasonal movements between upland and lowland pastures, such as practiced in Norway, Scotland, and Switzerland, as "vertical transhumance." Nomadism, in which entire human groups migrate with their livestock, has not generally characterized livestock raising in western and central Europe since the Neolithic.

PERIPHERY. World-systems theorists use this term to designate societies in a position of economic subservience to a *core* area, often subject to resource extraction and disadvantageous trading relationships. If incorporated into a political empire, peripheral areas may be required to pay taxes in various forms to the core authority. Some of the northern European regions that were among the peripheries of the Roman Empire became core areas in the sixteenth and seventeenth centuries as they established commercial and political empires in the Americas, followed by imperial conquest in Africa and southeast Asia in the nineteenth century.

PHYLLOXERA. A disease affecting the rootstock of grapevines, native to North America and caused by aphidlike insects. Attacking European vines after 1860, phylloxera decimated grapevines throughout much of France and constituted the worst ecological disaster of the nineteenth century from the perspective of French winegrowers. Phylloxera was stemmed through hybridization with resistant species and, more commonly, grafting grapevines onto resistant rootstock from North America.

PHYSIOCRATS. Eighteenth-century French economists influenced by Enlightenment thought, particularly the emphasis on individual liberties. They were among the first theorists of the free market and invented the concept of laissez-faire. Physiocrats believed that a nation's prosperity was tantamount to its agricultural productivity, and they advocated a series of measures to heighten it, from abolishing controls on the grain trade, to equalizing taxation and plowing up all available land. Their policies influenced the French monarchy's financing of land reclamation in the mid-eighteenth century, yet they eventually came into conflict with foresters, who viewed forests as much more than simply "uncultivated" land fit to be logged.

PLAN MESSMER. Announced in 1974 and named after Pierre Messmer, the prime minister of France at the time, the Plan aimed at a vast increase in France's capacity to generate nuclear energy. Implementing the Plan reduced France's dependency on imported petroleum but greatly added to the country's foreign debt. Four-fifths of France's electricity now comes from nuclear power plants, with much left over for export, making France a uniquely nuclear country within Europe. The national electricity utility (EDF) stores its spent fuel underground following vitrification, a method considered by some scientists to represent only a temporary solution to handling highly toxic radioactive waste.

PLEISTOCENE. The second geologic epoch of the Quaternary period, lasting from 2.3 million years ago until 10,000 years ago and marked by successive glaciations and interglacials.

PODZOLIZATION. The formation of acidic soil through the leaching of minerals from upper to lower layers. The light, sandy soils of west Jutland (Denmark) became podzolic following deforestation and agriculture during the early Iron Age. Precipitation washed out minerals from these fields, which had been demoted to pasture by the end of the Iron Age. A similar process explains the formation of lowland heaths in Britain.

POSITIVE FEEDBACK. The acceleration of a process in an open system such that a new or altered variable changes even more in the same direction. The term can be applied to engineering and economics as well as to ecology. Positive feedback characterized European manufacturing during the Industrial Revolution, as the surplus energy derived from coal was fed back into the productive process, thus accelerating economic growth and the use of natural resources.

PRECAUTIONARY PRINCIPLE. This concept, strongly linked to that of "prevention," provides a framework for environmental policy by advising against the use of certain technologies that carry a high risk of negative environmental consequences. By requiring proof of the absence of negative consequences, the precautionary principle would, if fully implemented, sharply alter the historical trajectory of technological development. In the developed world, industry has generally applied new technologies in a laissez-faire regime, hampered only by specific regulations and laws informed by a proof, or near certainty, of negative consequences stemming from use of the technology. The precautionary principle thus would shift the burden of proof in all policymaking and jurisprudence related to technology. Hans Jonas, a German philosopher, formulated the principle in his 1979 book *The Imperative of Responsibility*. The 150 signatories to the Rio Declaration, one formal outcome of the United Nations Conference on Environment and Development, theoretically adopted it in 1992. France has institutionalized the precautionary principle through the Barnier Law of 1995, and many Europeans feel strongly about applying it to the risks associated with genetically modified organisms.

PRICE REVOLUTION. From the later fifteenth century to the mid-seventeenth century, high inflation affected the economy of western Europe, producing a sixfold rise in prices over approximately 150 years. The vast increase in central European silver production caused the initial price rise, and inflation was subsequently fueled by the influx of gold and silver bullion from the New World. Demographic pressure from the late fifteenth century also put pressure on available agricultural resources and hence prices. Boosted by the inflow of silver from Latin America and greater demand, rising prices stimulated farmers to cultivate more intensively and to put more land under the plow.

PROTOINDUSTRY. Also known as the putting-out system, protoindustry describes nonmechanized manufacturing (most often of textiles) carried out in rural households, often by part-time laborers. Merchant manufacturers supply laborers with raw materials, purchase the finished items, and transport them to towns either for finishing or sale. In the early modern centuries protoindustry carried much potential for growth and characterized many of the pioneering regions of Europe's later Industrial Revolution.

REMEMBREMENT. The process, in France, of consolidating dispersed, small plots into larger farms by means of communal redistribution. Encouraged by officials in the French government, *remembrement* was one of various policies that helped create a highly modernized agriculture requiring relatively few farmers. *Remembrement* and like trends led to a 70 percent decline in the French farming population between the 1950s and the 1990s. The phenomenon was most visible in those parts of France formerly marked by hedgerows, namely Brittany. As occurred in England, hedgerows were either removed completely or thinned.

RUDORFF, ERNST (1840–1916). Rudorff was a professional composer and musicologist who became an influential theorist and actor in Germany's early conservation movement. In his founding manifesto for the German League for the Protection of the Homeland (1904), Rudorff railed against the industrial exploitation of nature, scientific forestry, and modern materialism in the broadest sense. Rudorff typified the popular *völkisch* idea that German nature was intimately bound with German identity; to destroy the first was to damage the second. Thus, the preservation of rural and domesticated landscapes, as well as more wild regions, should be, in his view, an urgent national priority. Partial to the sublime aesthetic, Rudorff went as far as castigating tourism in his essay "On the Relationship of Modern Life to Nature" (1880). He complained that, far from providing possible recruits to the cause of conservation, most nature tourism produced blights on picturesque landscapes in the form of extended rail lines and hotels. Rudorff represented a radical Romantic voice within Germany's diverse *Naturschutz* (nature protection) movement.

RUSKIN, JOHN (1819–1900). Best known as an author and art critic, Ruskin devoted much professional energy to the preservation of Britain's natural and cultural heritage. He was a vocal and prestigious supporter of the preservation of Lake Thirlmere in the 1870s, the first major preservationist battle in nineteenth-century Britain. Ruskin helped inspire the formation of Britain's National Trust for Places of Historic Interest or Natural Beauty in 1895.

SAINT FRANCIS OF ASSISI (1182–1226). A Roman Catholic spiritual reformer and founder of the Franciscan order of friars, who followed monastic rules of asceticism but did not withdraw from the world, preaching, begging, and serving as living examples of charity. Much later declared a patron saint of animals and the environment, Saint Francis preached and wrote of the kinship he felt with all beings, animate and inanimate. His canticle *Laudes creaturarum* contains his invocations of Brother Sun and Sister Moon, and contemporary stories tell of his speaking with birds and taming a ferocious wolf.

SCIENTIFIC REVOLUTION. A sea change in Western thought, beginning in the sixteenth century, that sought to apply reason and empiricism to studying the natural world, including the human body. The Scientific Revolution is most famously associated with advances in astronomy and physics, namely the rejection of the Ptolemaic, or geocentric, model of the solar system, long adopted by the Roman Catholic Church, in favor of the heliocentric, or sun-centered, model theorized by Nicolaus Copernicus (1473–1543). Its basic methodological contribution was the scientific method itself—the use of experimentation, observation, and deductive reasoning—in the discovery of scientific "truths." Englishman Sir Francis Bacon (1561–1626) vigorously advocated both the scientific method and a mechanistic understanding of nature, as did mathematician René Descartes (1596–1650). Mechanism assumed that nature consisted only of matter and strictly obeyed physical laws, a view that supplanted common earlier understandings that attributed spirit or intelligence to a living nature and assumed the role of magical forces.

SMOG. English physician Dr. H. A. Voeux coined the term in 1905 to indicate the combination of smoke and fog that had plagued London, above all other European cities, with increasing frequency in the nineteenth century. Caused primarily by the domestic burning of coal, London's smog occasionally caused excess mortality, as in 1873 and 1952. From the later twentieth century onward, smog was mostly associated with the haze produced by specific photochemical reactions: emissions of hydrocarbons and nitrogen oxides from automobile tailpipes combine with sunlight to form pollutants such as ozone. This form of smog has plagued many northern European cities but is worst in sunny cities at more southern latitudes that are located in geographic basins, for example, Los Angeles and Mexico City.

SOIL HORIZONS. Soil forms distinct layers characterized by similar color and structure. Typically, soils show the following profile marked by three horizons: an A-horizon nearest the surface, which contains most of the organic matter found in soil; a B-horizon, where iron or aluminum accumulate; and a C-horizon of undisturbed mineral material. *Podzolization* produces an E-horizon leached of minerals and humus.

SUBLIME. A quality often attributed to certain landscapes by early nineteenth-century Romantic writers and artists. Emphasizing the juxtaposition of pleasure with danger and mystery, sensations derived from the perception of, for example, alpine crags or thundering waterfalls, the sublime aesthetic greatly influenced both European and North American preservationists as they selected areas deemed worthy of preservation. In his essay "On the Sublime," German poet Friedrich Schiller suggested the connections between the experience of nature and creative thought and action: "'Who knows how many luminous thoughts or heroic resolves, which no saloon or student's cell would have given the world, have sprung from the valorous conflict of the mind with the great spirit of Nature in a single walk?'" (Dominick 1992, 25)

SUBSISTENCE. The action or means of producing what is necessary for one's own survival. A mode of production characterized by subsistence means there is an absence of surplus, and thus of trade; pure subsistence, however, is rarely identified among historical human groups. Activities that can be engaged in to assure subsistence include hunting and gathering, fishing and trapping, horticulture, agriculture, pastoralism, and agropastoralism.

SUSTAINABLE DEVELOPMENT. The concept was first launched in the World Conservation Strategy of the *International Union for the Conservation of Nature and Natural Resources.* It received its enduring definition in the 1987 report of the *Brundtland Commission,* as follows: "to ensure that [development] meets the needs of the present without compromising the ability of future generations to meet their own needs." The concept had wide appeal and was adopted by international organizations, particularly the United Nations, in order to build concrete links between environmental protection and economic development. Initially a broadly conceived set of orientations for technology, investment, and resource use, it has since worked its way into many instances of policymaking at the level of the nation-state (particularly in Norway) as well as the European Union. See the case study "From Nature Conservation to Sustainable Development: The Scandinavian Experience" on pages 188–203.

TECHNOCRAT. A member of a socially elite group of technicians, often trained in engineering or economics, enlisted to manage the economy and run the government. Values associated with technocracy were spawned in the United States in the early twentieth century and gained a significant audience in post–World War II Europe. Technocrats helped shape the major economic trends and devise the large-scale technological projects that have had profound social and environmental impacts in Europe during the last half century.

TRAGEDY OF THE COMMONS. Garrett Hardin popularized this expression in an article appearing in *Science* in 1968. As a metaphor or model, it expresses the degradation or destruction of public goods, "commons," through overuse by individuals or private interests with unlimited rights of use. It assumes that the latter will rationally choose to exploit more than their rightful share of the public good, ignoring the social and environmental costs of doing so because individual interests are being maximized at little perceptible cost. As a description of historical practices, the concept may have most relevance to modern instances of large-scale pollution and overexploitation, namely of "common" rivers, forests, and oceans. Perhaps the best example is the collapse of fisheries in the world's oceans, a truly global resource under no single system of stewardship. The metaphor applies less well to the historical use of the commons in the sense of agrarian communal property or common use rights. See *common access*.

TRANSHUMANCE. In the panoply of ways to raise livestock (see *pastoralism*), transhumance refers to the long-distance migrations of often very large herds of sheep or cattle between summer and winter pastures. In contrast to pastoralism, transhumance takes place between relatively distant regions, as between Provence (winter) and the southern Alps (summer). Professional shepherds generally accompanied the herds along well-defined migration routes. Whereas pastoralism usually had a communal or familial basis, transhumance occurred within a larger commercial context. In the central Alps of Austria and Switzerland this was a new adaptation of the thirteenth and later medieval centuries, first pushing the margin of use upward, then making use of strong late medieval markets for butter and cheese. In the latter case it paralleled different agro-ecosystems then developed in Denmark and the northern Low Countries. Transhumance to and from the Alps was practiced until well into the nineteenth century. Transhumant herds numbering in the thousands could cause severe erosion along their paths, but the practice enabled domestic animals and their owners to use seasonally available grazing resources.

USUFRUCT. The right to use property belonging to someone else, as long as damage or alteration to the property does not occur. Also known as use rights. See *common access.*

WAVE OF ADVANCE. An influential model used to explain the progression of agriculture across Europe in the course of the Neolithic. Proposed in 1973 by archaeologist Albert Ammerman and geneticist Luca Cavalli-Sforza, the model attempts to account for this progression in terms of population growth and migration. Based on the examination of radiocarbon dates of Neolithic sites, Ammerman and Cavalli-Sforza estimated that the Neolithic way of life was carried by migrants in the general direction of southeast to northwest, moving across Europe at an average rate of one kilometer per year. In their view, Europe was colonized by farming peoples emanating from Anatolia and the Balkans. Subsequent research by other specialists has cast doubt on the possibility of colonization in parts of the North European Plain, Scandinavia, the Atlantic littoral, and the British Isles, suggesting instead a process of indigenous acculturation. More complex models emphasize that degrees of migration and acculturation may not be mutually exclusive when examining specific regions, such as the British Isles. See *Neolithic package.*

WILDWOOD. Forest historian Oliver Rackham employs this term to describe the prehistoric forests of Great Britain. Developing from the end of the last glaciation, these forests produced climax plant communities in the sixth and fifth millennia BCE. A diverse, highly regionalized forest, the wildwood was dominated by pine and birch in the Scottish Highlands and some mountains of Ireland, hazel and elm in most of Ireland and southwestern England, lime in the lowlands of England, and alder throughout Britain in proximity to lakes. Early Britons modified the wildwood, particularly through fire, during both the Paleolithic and Mesolithic eras, but during the Neolithic, Bronze, and Iron Ages Britons accelerated the destruction of wildwood. This general forest history applies as well to continental Europe, though a severe retreat of the forest in the Scottish Highlands occurred as early as the Bronze Age under the dual impacts of pastoralism and a wetter climate that impeded regeneration.

BIBLIOGRAPHIC ESSAY

General introductory volumes should be mentioned before we turn to northern Europe. Clive Ponting's well-known *A Green History of the World* (London: Sinclair-Stevenson, 1991) makes good reading but is surpassed by the more sophisticated, though rather technical, study of Ian G. Simmons, *Changing the Face of the Earth: Culture, Environment, History* (Oxford: Blackwell, 1989). Two additional works treating global environmental history are Sing C. Chew, *World Ecological Degradation: Accumulation, Urbanization, and Deforestation 3000 BC–AD 2000* (Walnut Creek, CA: Alta Mira Press, 2001) and Joachim Radkau, *Natur und Macht. Eine Weltgeschichte der Umwelt* (Munich: C. H. Beck, 2000), which frames both the exploitation and protection of nature within a history of state-building. Also relevant is William H. McNeill's *Plagues and Peoples* (New York: Anchor Books, 1989), a wide-ranging study of disease in history. Some nice reflections on the subject of environmental history can be found in a Finnish booklet edited by Timo Myllyntaus and Mikko Saikku, *Encountering the Past in Nature: Essays in Environmental History* (Helsinki: Helsinki University Press, 1999).

Useful introductions to the environmental history of northern Europe include Robert Delort and François Walter, *Histoire de l'environnement européen* (Paris: Presses Universitaires de France, 2001); I. G. Simmons, *An Environmental History of Great Britain* (Edinburgh: Edinburgh University Press, 2001); and *Pour une histoire de l'environnement*, ed. Corinne Beck and Robert Delort (Paris: CNRS Editions, 1993). An important collection of articles can be found in the volume edited by Peter Brimblecombe and Christian Pfister, *The Silent Countdown: Essays in European Environmental History* (Berlin: Springer-Verlag, 1990). A somewhat older and idiosyncratic book, but still worth reading regarding the environmental history of central Europe, is Helmut Jäger, *Einführung in die Umweltgeschichte* (Darmstadt: Wissenschaftliche Buchgesellschaft, 1994). More recent is Rolf Peter Sieferle's *Rückblick auf die Natur: Eine Geschichte des Menschen und seiner Umwelt* (Munich: Luchterhand, 1997). Stephen J. Pyne offers a sweeping account of European fire in *Vestal Fire: An Environmental History, Told through Fire, of Europe and Europe's Encounter with the World* (Seattle: University of Washington Press, 1997). Readers pursuing the history of western attitudes toward nature should turn to the somewhat outdated but still highly rewarding work of Clarence Glacken, *Traces on the Rhodian Shore: Nature and Culture in Western Thought from Ancient Times to the End of the Eighteenth Century* (Berkeley: University of California Press, 1967) and the more recent book by Peter A. Coates, *Nature: Western Attitudes since Ancient Times* (Berkeley: University of California Press, 1998).

Environmental History and *Environment and History* are the field's leading journals, and readers should also consult the bibliography available on the website http://www.eseh.org of the European Society for Environmental History. See also the recent *Dealing with Diversity: Proceedings of the Second International Conference of the European Society for Environmental History* (Prague: Charles University, 2003). *Environmental History* has published historiographical essays on various European nations: see Mark Cioc et al., "Environmental History Writing in Northern Europe," *Environmental History* 5, no. 3 (July 2000): 396–406; and Mark Cioc et al., "Environmental History Writing in Southern Europe," *Environmental History* 5, no. 4 (October 2000): 545–556. Readers can also find relevant historiographies posted on the internet at http://www.h-net.org/~environ/historiography/. For example, see Timo Myllyntaus, "Writing about the Past in Green Ink: The Emergence of Finnish Environmental History," http://www.h-net.org/~environ/historiography/finland.htm and Matt Osborn, "Sowing the Field of British Environmental History," http://www.h-net.org/~environ/historiography/british.htm, which includes a discussion of France.

Historical geographies remain important sources for the study of European environmental history. Consult R. A. Butlin and R. A. Dodgshon, *An Historical Geography of Europe* (Oxford: Clarendon, 1998); Xavier Planhol and Paul Claval, *An Historical Geography of France*, trans. Janet Lloyd (Cambridge, UK: Cambridge University Press, 1994); and *Themes in the Historical Geography of France*, ed. Hugh D. Clout (London: Academic Press, 1977).

PALEOLITHIC AND MESOLITHIC ERAS

Historians must be prepared to spread their wings when delving into the Paleolithic, Mesolithic, and Neolithic eras of European history. Paleoanthropology and related fields have long confronted issues of concern to environmental historians, namely human adaptations to and effects upon evolving ecosystems. A recommended starting place is Randall White's *Dark Caves, Bright Visions: Life in Ice Age Europe* (New York: American Museum of Natural History and W. W. Norton & Co., 1986), which accompanies the exhibition of Paleolithic art that took place at the American Museum of Natural History, in collaboration with the Musée des Antiquités Nationales and the Musée de l'Homme in France. The volume provides an excellent, lucid introduction for the nonspecialist to the survival strategies, social organization, and, especially, toolmaking and artistic endeavors of modern humans during the various cultural periods of the Upper Paleolithic. Stunning photographs from the exhibition are a highlight of the volume.

In *The Human Career: Human Biological and Cultural Origins*, 2d ed. (Chicago: University of Chicago Press, 1999), Richard G. Klein provides a lengthy but highly readable and well-illustrated guide to primate and specifically human evolution for the nonspecialist. Later chapters explore the morphology, survival strategies, technologies, and demography of Neanderthals and anatomically modern humans. The book concludes with a bibliography including over 2,500 sources. Devoted to the Neanderthals, an equally readable and comprehensive text is Paul Mellars, *The Neanderthal Legacy: An Archaeological Perspective from Western Europe* (Princeton, NJ: Princeton University Press, 1996).

No study of the European Paleolithic should neglect Clive Gamble, *The Paleolithic Societies of Europe* (Cambridge, UK: Cambridge University Press, 1999). This scholarly text is not recommended for the novice, however, and Gamble's central argument concerns the social organization of Paleolithic Europeans.

Several essays in Geoff Bailey's edited volume *Hunter-Gatherer Economy in Prehistory: A European Perspective* (Cambridge: Cambridge University Press, 1983) are useful to the study of northern Europe. See Marcie Madden, "Social Network Systems amongst Hunter-Gatherers Considered within Southern Norway," 191–200, for insight into the environmental context of subsistence during the Norwegian Mesolithic; Clive Gamble, "Culture and Society in the Upper Palaeolithic of Europe," 201–211, for more general information about climate and social change; and Michael A. Jochim, "Palaeolithic Cave Art in Ecological Perspective," 212–219, though it has greater relevance to the study of Franco-Cantabria.

Questions surrounding the timing and ecological circumstances of the first human occupation of Europe are elucidated by the essays in *The Earliest Occupation of Europe: Proceedings of the European Science Foundation Workshop at Tautavel (France), 1993*, ed. Wil Roebroeks and Thijs van Kolfschoten (University of Leiden: Analecta Praehistorica Leidensia 27, 1995). The volume includes Clive Gamble's essay, "The Earliest Occupation of Europe: The Environmental Background," 279–295.

Many issues pertinent to the Upper Paleolithic are treated by Robin Dennell in *European Economic Prehistory: A New Approach* (London: Academic Press,

1985); especially useful are Dennell's discussion of adaptations during the last glacial maximum and analysis of the Mesolithic in comparison to the Neolithic.

The authors most devoted to reconstructing Paleolithic landscapes are Tjeerd H. van Andel and P. C. Tzedakis. See the following articles: "Priority and Opportunity: Reconstructing the European Middle Paleolithic Climate and Landscape," in *Science in Archaeology: An Agenda for the Future*, ed. Justine Bayley (London: English Heritage, 1998), 37–45; and "Paleolithic Landscapes of Europe and Environs, 150,000–25,000 Years Ago: An Overview," *Quaternary Science Reviews* 15 (1996): 481–500.

A comprehensive multiauthor guide to the faunal extinctions of the Quaternary period on all continents is *Quaternary Extinctions: A Prehistoric Revolution*, ed. Paul S. Martin and Richard G. Klein (Tucson: University of Arizona Press, 1984).

A more regionally focused approach can be found in *Studies in the Upper Palaeolithic of Britain and Northwest Europe*, ed. Derek A. Roe (Oxford: BAR International Series 296, 1986), in particular the essays by John R. Campbell and Katharine Scott.

NEOLITHIC ERA THROUGH THE ROMAN EMPIRE

An excellent point of departure for the study of the European Neolithic is Alasdair Whittle's "The First Farmers," chap. 4 of *The Oxford Illustrated Prehistory of Europe*, ed. Barry Cunliffe (Oxford: Oxford University Press, 1994). The essays edited by T. Douglas Price in *Europe's First Farmers* (Cambridge: Cambridge University Press, 2000) provide the best overall guide to current scholarship in the study of early agriculture in Europe; the volume covers all regions of the continent. I. J. Thorpe summarizes much scholarship on the Neolithic in *The Origins of Agriculture in Europe* (London and New York: Routledge, 1996). The classic argument for the role of population pressure in the origins of agriculture is that of Mark N. Cohen, *The Food Crisis in Prehistory: Overpopulation and the Origins of Agriculture* (New Haven, CT: Yale University Press, 1977). Among other journals, *Antiquity* has been one forum for recent debates about the European Neolithic. See the article by Marek Zvelebil, "On the Transition to Farming in Europe, or What Was Spreading with the Neolithic: A Reply to Ammerman," *Antiquity* 63 (1989): 379–383.

Regional approaches are imperative for a fuller understanding of the complexities of the European Neolithic. See, for example, Peter Bogucki's *Forest Farmers and Stockherders: Early Agriculture and Its Consequences in North-Central Europe* (Cambridge: Cambridge University Press, 1988); Leendert P.

Louwe Kooijmans, "The Mesolithic/Neolithic Transformation in the Lower Rhine Basin," in *Case Studies in European Prehistory*, ed. Peter Bogucki (Boca Raton, FL: CRC Press, 1993), 95–143; and the following essays in *Harvesting the Sea, Farming the Forest: The Emergence of Neolithic Societies in the Baltic Region*, ed. Marek Zvelebil, Lucyna Domanska, and Robin Dennell (Sheffield: Sheffield Academic Press, 1998): Marek Zvelebil, Lucyna Domanska, and Robin Dennell, "Introduction: The Baltic and the Transition to Farming," 1–7; Marek Zvelebil, "Agricultural Frontiers, Neolithic Origins, and the Transition to Farming in the Baltic Basin," 9–27; Kristina Jennbert, "'From the Inside': A Contribution to the Debate about the Introduction of Agriculture in Southern Scandinavia," 31–35; Leonid Zaliznyak, "The Ethnographic Record and Structural Changes in the Prehistoric Hunter-Gatherer Economy of Boreal Europe," 45–50; and Peter Bogucki, "Holocene Climatic Variability and Early Agriculture in Temperate Europe: The Case of Northern Poland," 77–85. Disparate introductions to the British Neolithic can be found in Alasdair Whittle, "The Coming of Agriculture: People, Landscapes, and Change c. 4000–1500 BC," in *The Peopling of Britain: The Shaping of a Human Landscape*, ed. Paul Slack and Ryk Ward (Oxford: Oxford University Press, 2002), 77–109 [noted in text as Whittle 2002a]; chap. 3 of I. G. Simmons, *An Environmental History of Great Britain: From 10,000 Years Ago to the Present* (Edinburgh: Edinburgh University Press, 2001), whose account pays explicit attention to ecological changes due to agriculture; and Ian Armit and Bill Finlayson's "Hunter-Gatherers Transformed: The Transition to Agriculture in Northern and Western Europe," *Antiquity* 66 (1992): 664–676. The French Neolithic is well treated in Christopher Scarre's edited volume titled *Ancient France: Neolithic Societies and Their Landscapes, 6000–2000 BC* (Edinburgh: Edinburgh University Press, 1983).

For views of ancient forests and woodmanship in a British context, see Oliver Rackham, "Trees and Woodland in a Crowded Landscape—The Cultural Landscape of the British Isles," in *The Cultural Landscape: Past, Present and Future*, ed. Hilary H. Birks, H.J.B. Birks, Peter Emil Kaland, and Dagfinn Moe (Cambridge: Cambridge University Press, 1988), 53–77; and Rackham, *Ancient Woodland: Its History, Vegetation and Uses in England* (London: Edward Arnold, 1980). Megalithic monuments are analyzed with reference to landscape in *Monuments and Landscape in Atlantic Europe: Perception and Society during the Neolithic and Early Bronze Age*, ed. Christopher Scarre (London and New York: Routledge, 2002); see especially Scarre's introduction and Alasdair Whittle's conclusion.

Essential background on post-Neolithic European societies, economies, and political systems can be found in Sarunas Milisauskas, *European Prehistory* (New York: Academic Press, 1978). Barry Cunliffe's essays should be consulted for background on the Bronze Age, Iron Age, and Roman Europe; see his "Tribes and Empires c.1500 BC–AD 500" in *The Peopling of Britain* (see above); and also see his more global essays contained in chaps. 10 and 12 of *The Oxford Illustrated Prehistory of Europe* (see above). Agriculture and metallurgy receive excellent treatment in chap. 9 of Françoise Audouze and Olivier Büchsenschütz, *Towns, Villages and Countryside of Celtic Europe* (London: B. T. Batsford, 1991). Though largely devoted to the origins of the state, Lotte Hedeager's *Iron-Age Societies: From Tribe to State in Northern Europe, 500 BC to AD 700* (Oxford: Blackwell Publishers, 1992) contains valuable information on economy and subsistence. See also the following essays in *Science in Archaeology: An Agenda for the Future,* ed. Justine Bayley (London: English Heritage, 1998): Justine Bayley, "Metals and Metalworking in the First Millennium AD," 161–168; and Marijke van der Veen and Terry O'Connor, "The Expansion of Agricultural Production in Late Iron Age and Roman Britain," 127–143.

Works by Peter Wells provide vital perspectives on the peoples inhabiting the northern provinces as well as frontier zones of the Roman Empire. See his books *The Barbarians Speak: How the Conquered Peoples Shaped Roman Europe* (Princeton, NJ: Princeton University Press, 1999), and *Beyond Celts, Germans and Scythians* (London: Duckworth, 2001), especially chap. 5. J. Donald Hughes's *Pan's Travail: Environmental Problems of the Ancient Greeks and Romans* (Baltimore and London: Johns Hopkins University Press, 1994) contains limited information about northern Europe but offers a global analysis of the ecological impacts of the Roman Empire, as does chap. 5 of Sing C. Chew's *World Ecological Degradation* (see above). Stephen Rippon's *The Transformation of Coastal Wetlands: Exploitation and Management of Marshland Landscapes in North West Europe during the Roman and Medieval Periods* (Oxford: Oxford University Press, 2000) is a highly informative study of the human occupation and transformation of "marginal" landscapes over a significant period of time.

MIDDLE AGES

Further general reading on relations between humans and their environment in medieval Europe is limited by the realities of medievalist scholarship in this innovative field. Much essential information has been assembled by scholars who never conceived of doing environmental history, or even of working within an ecological paradigm. This is especially true of much agricultural history. The best and most self-conscious new research is still more often found in scholarly journals and collective volumes rather than monographs.

A strong assertion by Lynn White that medieval western Christianity was responsible for later environmental degradation provoked thoughtless repetition and widespread vigorous rebuttal. See Lynn T. White Jr., "The Historical Roots of Our Ecologic Crisis." in his *Dynamo and Virgin Reconsidered*, 75–94 (Cambridge, MA: MIT Press, 1968). Some examples of rebuttal include R. Attfield, "Christian Attitudes to Nature," *Journal of the History of Ideas* 44 (1983): 369–386; E. Whitney, "Lynn White, Ecotheology, and History," *Environmental Ethics* 15 (1993): 151–169; and Manussos Marangudakis, "The Medieval Roots of Our Ecological Crisis," *Environmental Ethics* 23 (2001): 243–260.

Other scholars approach related issues as problems in intellectual and cultural history: David Herlihy, "Attitudes toward the Environment in Medieval Society," in L. Bilsky, ed., *Historical Ecology: Essays on Environment and Social Change* (Port Washington, NY: Kinnikat Press, 1980), 100–116; Jeremy Cohen, *"Be fertile and increase, fill the earth and master it": The Ancient and Medieval Career of a Biblical Text* (Ithaca, NY: Cornell University Press, 1989); Roger Sorrell, *St. Francis of Assisi and Nature: Tradition and Innovation in Western Christian Attitudes toward the Environment* (New York: Oxford University Press, 1988); Stephen Wilson, *The Magical Universe: Everyday Ritual and Magic in Pre-Modern Europe* (London and New York: Hambledon, 2000); and Peter Biller, *The Measure of Multitude: Population in Medieval Thought* (Oxford: Oxford University Press, 2001).

Quite a few general or regional books treat medieval environments in the context of everyday material life and the history of technology, including agriculture. See Norman Pounds, *Hearth and Home: A History of Material Culture* (Bloomington: Indiana University Press, 1989); Lynn T. White Jr., *Medieval Technology and Social Change* (Oxford: Oxford University Press, 1962); an original, seminal, and much criticized work, Jean Gimpel, *The Medieval Machine: The Industrial Revolution of the Middle Ages* (New York: Holt, Rinehart and Winston, 1976). Also see Elizabeth B. Smith and Michael Wolfe, eds., *Technology and Resource Use in Medieval Europe: Cathedrals, Mills and Mines* (London: Ashgate, 1997); W. Groenman-van Waateringe and L. H. van Wijngaarden-Bakker, eds., *Farm Life in a Carolingian Village* (Assen/Maastricht, Netherlands, and Wolfeboro, NH: Van Gorcum, 1987); Del Sweeney, ed., *Agriculture in the Middle Ages: Technology, Practice, and Representation* (Philadelphia: University of Pennsylvania Press, 1995); Bruce M. S. Campbell, *English Seigniorial Agriculture 1250–1450* (Cambridge: Cambridge University Press, 2000).

Woodland studies offer some contrasts of regions, periods, and methodologies. See Oliver Rackham, *Trees and Woodland in the British Landscape*, rev. ed. (London: Dent, 1993), for a debunking of mythic clichés; Chris Wickham, "European Forests in the Early Middle Ages: Landscape and Land Clearance," in his *Land and Power: Studies in Italian and European Social History 400–1200* (London: British School at Rome, 1994), 155–199; Roland Bechmann, *Trees and Man: The Forest in the Middle Ages* (New York: Paragon, 1990); R. Grant, *The Royal Forests of England* (London: Alan Sutton, 1991); Jean Birrell, "Common Rights in the Medieval Forest," *Past & Present* 117 (1987): 22–49; Michael Williams, *Deforesting the Earth: From Prehistory to Global Crisis* (Chicago: University of Chicago Press, 2002).

Monographs on climatic history and natural catastrophes should be checked against the latest research in scientific journals, for this is a rapidly evolving field. Good starting points include historical climatologist Jean Grove's *The Little Ice Age* (London: Methuen, 1988); M.G.L. Baillie, "Putting Abrupt Environmental Change Back into Human History," in *Environments and Historical Change: The Linacre Lectures 1998*, ed. Paul Slack (Oxford: Oxford University Press, 1999), 46–75; *The Years without Summer: Tracing AD 536 and Its Aftermath*, ed. J. D. Gunn (Oxford: Archaeopress, 2000), 5–204; Brian Fagan, *The Little Ice Age: How Climate Made History, 1300–1850* (New York: Basic Books, 2000), the work of an archaeologist writing for a general audience.

Relations among human populations, parasitic diseases, and nutrition are best accessed through recent and revisionist studies on the demographic crises of the later Middle Ages, such as William C. Jordan, *The Great Famine: Northern Europe in the Early Fourteenth Century* (Princeton, NJ: Princeton University Press, 1996); Norman F. Cantor, *In the Wake of the Plague: The Black Death and the World It Made* (New York: Free Press, 2001); Samuel Kline Cohn, *The Black Death Transformed: Disease and Culture in Early Renaissance Europe* (London: Arnold, 2002).

Regional landscape studies integrate various themes from both biogeographical and cultural perspectives. See *People and Nature in Historical Perspective*, ed. Jozsef Laszlovsky and Peter Szabo (Budapest: Central European University Department of Medieval Studies and Archaeolingua, 2003); *Inventing Medieval Landscapes: Senses of Place in Western Europe*, ed. J. Howe and M. Wolfe (Gainesville: University Press of Florida, 2002); *The Medieval and Early-Modern Rural Landscape of Europe under the Impact of the Commercial Economy*, ed. H. J. Nitz (Göttingen: Geographical Institute of the University of Göttingen, 1987); Petra Dark, *The Environment of Britain in the First Millennium AD* (London: Duckworth, 2000); William TeBrake, *Medieval Frontier: Culture and Ecology in Rijnland* (College Station: Texas AM University Press, 1985).

A sampler of works on humans and animals is Nicholas Orme, "Medieval Hunting, Fact and Fancy," in *Chaucer's England: Literature in Historical Context*, ed. Barbara A. Hanawalt (Minneapolis: University of Minnesota Press,

1992), 133–153; Joyce Salisbury, *The Beast Within: Animals in the Middle Ages* (New York: Routledge, 1994); Kathleen Biddick, *The Other Economy: Pastoral Husbandry on a Medieval Estate* (Berkeley and Los Angeles: University of California Press, 1989); John Langdon, *Horses, Oxen and Technological Innovation. The Use of Draught Animals in English Farming from 1066–1500* (Cambridge: Cambridge University Press, 1986).

Most works on water take a technological perspective, although the first listed below also has ecological considerations: *Working with Water in Medieval Europe: Technology and Resource Use,* ed. Paolo Squatriri (Leiden: Brill, 2000); André E. Guillerme, *The Age of Water: The Urban Environment in the North of France, A.D. 300–1800* (College Station: Texas A & M University Press, 1988); Roberta J. Magnusson, *Water Technology in the Middle Ages: Cities, Monasteries, and Waterworks after the Roman Empire* (Baltimore: Johns Hopkins University Press, 2001).

EARLY MODERN ERA

One dominating theme in the historiography of early modern (northwest) Europe is the environmental background of the intertwined processes of economic growth, geographical expansion, and agricultural modernization. The Industrial Revolution often figures as an "outcome" of these early modern developments, though we have chosen to treat it separately (see pertinent bibliography below). A good statistical study of land productivity in northern Europe is edited by the Flemish historians Bas J. P. van Bavel and Erik Thoen, *Land Productivity and Agro-systems in the North Sea Area: Middle Ages–20th Century, Elements for Comparison* (Turnhout: Brepols, CORN Publication Series, 1999), which also contains further information about (and a confirmation of) the studies made by Bernhard H. Slicher van Bath on yield ratios in the 1960s and 1970s; see, for example, "The Yields of Different Crops, Mainly Cereals in Relation to the Seed c. 810–1820," *Acta Historiae Neerlandica* 2 (1967): 78–97.

The importance of "ghost acreage" for European economic development is central in John F. Richards, *The Unending Frontier: An Environmental History of the Early Modern World* (Berkeley: University of California Press, 2003). This book contains extended chapters on hunting, fishing, and whaling as well as on the "energy transformation" in England. (See the section on the modern era, below, for more on ghost acreage.)

Another central theme is the history of deforestation in Europe. This has proven to be a subject touching sensitive chords, for deforestation is directly related to questions of nature conservation and biodiversity. One should check the website of the Forest History Society. http://www.lib.duke.edu/forest/ Though not especially concerned with Europe, Andrew P. Dobson provides an excellent general introduction in *Conservation and Biodiversity* (New York: Scientific American Library, 1996). John Perlin's *A Forest Journey: The Role of Wood in the Development of Civilization* (New York: W. W. Norton, 1989) is highly readable but less sophisticated. Ian G. Simmons, *An Environmental History of Great Britain* (see above), however, is a must. For Scotland, see two volumes edited by T. C. Smout: *Scottish Woodland History* (Edinburgh: Scottish Cultural Press, 1997) and *Scotland since Prehistory: Natural Change and Human Impact* (Aberdeen: Scottish Cultural Press, 1993). Thorkild Kjaergaard's *The Danish Revolution, 1500–1800: An Ecohistorical Interpretation*, trans. David Hohnen (Cambridge and New York: Cambridge University Press, 1994) is a stimulating though contested study about politics, agriculture, deforestation, and ecological reconstruction. For studies of French forests, industry, and agriculture in the English language, one can turn to the historical geographies of Hugh D. Clout and Xavier de Planhol, mentioned at the beginning of this essay. On Germany see Paul Warde, "Forests, Energy, and Politics in the Early Modern German States," in *Il significatio dell'energia per la societa dal XIII al XVIII secolo*, ed. S. Cavaciocchi (Prato, Italy: Istituto Internazionale di Storia Economica "F. Datini," 2002). Many detailed but non-English-language studies pertaining to France, the Netherlands, Belgium, the German-speaking states, and Scandinavia exist. A good example is the Dutch study of Henk van Zon, *Geschiedenis en duurzame ontwikkeling. Duurzame ontwikkeling in historisch perspectief, enkele verkenningen* (Nijmegen, the Netherlands: SSN, 2002), an interesting study covering deforestation, mining, and the history of thinking about sustainability through the ages. It includes a summary in English.

For insights regarding early modern thinking about nature and the environment one can still use the well-written book of Keith Thomas, *Man and the Natural World: Changing Attitudes in England, 1500–1800* (London: Allen Lane, 1983). More recent is Peter A. Coates's *Nature: Western Attitudes since Ancient Times* (see above). On the "knowledge revolution" consult Joel Mokyr, *The Gifts of Athena: Historical Origins of the Knowledge Economy* (Princeton, NJ: Princeton University Press, 2002).

A third main theme treats early modern urbanization and urban pollution. Keith Thomas writes about it in the work cited above, and a classic is Peter Brimblecombe's *The Big Smoke: A History of Air Pollution in London since Medieval Times* (London and New York: Methuen, 1987). The most stimulating books are the French studies of André Guillerme, *The Age of Water: The Urban Environment in the North of France. A.D. 300–1800* (see above) and Alain Corbin, *The Foul and the Fragrant: Odor and the French Social Imagination*

(Cambridge, MA: Harvard University Press, 1986). Most of the numerous studies on urban pollution in Belgium and the Netherlands are published in Dutch. One should look at the *Jaarboek Ecologische Geschiedenis* (Yearbook of Ecological History) and its predecessor *Tijdschrift Ecologische Geschiedenis* (Journal of Ecological History), both published by Academia Press in Gent.

On climatic factors influencing adaptations in the early modern era, see the works on the Little Ice Age mentioned in the section on the Middle Ages in addition to the following: *Climate and History: Studies in Past Climates and Their Impact on Man*, ed. T.M.L. Wigley et al. (Cambridge, UK, and New York: Cambridge University Press, 1981) and H. H. Lamb, *Climate, History, and the Modern World*, 2d ed. (London and New York: Routledge, 1995). Climate history and the history of calamities, especially the Little Ice Age, have once more been thoroughly scrutinized by Christian Pfister and others in *Climatic Variability in Sixteenth-Century Europe and Its Social Dimension*, ed. Christian Pfister, Rudolf Brázdil, and Rüdiger Glaser (Dordrecht and Boston: Kluwer Academic Publishers, 1999). A recent treatment of the history of the Little Ice Age can be found in the book by John F. Richards (see above).

MODERN ERA

Modern history continues to receive the lion's share of attention from environmental historians, and the following recommendations represent a sampling of what is available.

A rare and fine study examining environmental dimensions of the French Revolution, in the context of small-scale peasant agriculture, is Peter McPhee's *Revolution and Environment in Southern France: Peasants, Lords, and Murder in the Corbières, 1780–1830* (Oxford: Oxford University Press, 1999).

Environmental perspectives on the causes of the English Industrial Revolution can be found in chap. 10 of Sidney Pollard, *Marginal Europe: The Contribution of Marginal Lands since the Middle Ages* (Oxford: Clarendon Press, 1997). Richard G. Wilkinson's seminal essay of 1973 on the industrial revolution as an ecological crisis is reprinted as "The English Industrial Revolution," in *The Ends of the Earth: Perspectives on Modern Environmental History*, ed. Donald Worster (Cambridge, UK: Cambridge University Press, 1988), 80–99. The central question concerns whether England or Europe in general would have been able to pursue economic growth without the transition to coal and the support of "ghost acreage." Georg Borgstrom introduced the term in *The Hungry Planet* (New York: Collier, 1972). Wilkinson did not use that phrase; however, Eric Lionel Jones applied it to history in 1981 in *The European Miracle: Environments,*

Economies, and Geopolitics in the History of Europe and Asia, 3d ed. (Cambridge, UK, and New York: Cambridge University Press, 2003). A profound successor is Kenneth Pomeranz, *The Great Divergence: China, Europe, and the Making of the Modern World Economy* (Princeton, NJ: Princeton University Press, 2000), which analyzes the possibility of Malthusian crisis in Europe at the end of the eighteenth century. Rolf Peter Sieferle also treats the "energy transformation" in *The Subterranean Forest: Energy Systems and the Industrial Revolution* (Cambridge: White Horse Press, 2001). For a local case study see Matthew Osborn, "'The Weirdest of All Undertakings': The Land and the Early Industrial Revolution in Oldham, England," *Environmental History* 8 (April 2003): 246–269.

An insightful look at the relative environmental costs of early industrialization is James Winter's *Secure from Rash Assault: Sustaining the Victorian Environment* (Berkeley: University of California Press, 1999). Both nineteenth- and twentieth-century developments are treated in *Le Démon moderne: la pollution dans les sociétés industrielles d'Europe,* ed. Christoph Bernhardt and Geneviève Massard-Guilbaud (Clermont-Ferrand: Presses Universitaires Blaise-Pascal, 2002). Works devoted to water pollution are numerous; in addition to specific essays in the edited volumes (see above), readers should consult the following studies: Jean-Pierre Goubert, *The Conquest of Water: The Advent of Health in the Industrial Age,* trans. Andrew Wilson (Cambridge, UK: Polity Press, 1989); Bill Luckin, *Pollution and Control: A Social History of the Thames in the Nineteenth Century* (Bristol and Boston: Adam Hilger, 1986); Dale H. Porter, *The Thames Embankment: Environment, Technology, and Society in Victorian London* (Akron, OH: University of Akron Press, 1998); and Donald Reid, *Paris Sewers and Sewermen: Realities and Representations* (Cambridge, MA: Harvard University Press, 1991). The many ways in which industrialization transformed the Rhine are powerfully treated in Mark Cioc, *The Rhine: An Eco-Biography, 1815–2000* (Seattle and London: University of Washington Press, 2002). Bent Jensen broadly covers environmental problems in modern Denmark with attention to the Danish urban environment in *Miljøproblemer og velfærd* (Copenhagen: Spektrum, 1996). A valuable primary source is Wilhelm Raabe's *Pfisters Mühle. Ein Sommerferienheft* (Leipzig, Germany: Grunow, 1884), one of the first environmentally inspired novels in Germany that tells the story of a miller whose livelihood is threatened by water pollution from a paper mill.

A leading work on urban air pollution is Peter Brimblecombe's *The Big Smoke* (see above). Franz-Josef Brüggemeier provides a detailed study of air pollution and related conflicts in nineteenth-century Germany in *Das unendliche Meer der Lüfte. Luftverschmutzung, Industrialisierung und Risikodebatten im*

19. Jahrhundert (Essen: Klartext, 1996). For a comparative perspective that treats American and German approaches to air pollution in the nineteenth and twentieth centuries, see Frank Uekötter, *Von der Rauchplage zur ökologischen Revolution. Eine Geschichte der Luftverschmutzung in Deutschland und den USA 1880–1970* (Essen: Klartext, 2003). Ole Hyldtoft treats industrial pollution in nineteenth-century Denmark in "Stank, kulrøg, og epidemier. Industri og miljø i Eanmark i 1800-årene," in the publication of the Twenty-First Nordic History Meeting held in Umeå in 1991, *Människan och miljön*, ed. Lars J. Lundgren (Umeå, Sweden: Historiska Institutionen, 1991), 115–141. Responses to both water and air pollution in Belgium are well covered in Christophe Verbruggen, "Nineteenth-Century Reactions to Industrial Pollution in Ghent, the Manchester of the Continent," in *Le Démon moderne* (see above). A Dutch parallel can be found in H. Diederiks and C. Jeurgens, "Environmental Policy in Nineteenth-Century Leyden," in *The Silent Countdown* (see above).

Readers should turn to the books by Winter and Smout (see above) for an introduction to agricultural modernization in Britain. The most comprehensive treatment of rural France over the past two centuries can be found in volumes 3 and 4 of *Histoire de la France rurale*, ed. Georges Duby and Armand Wallon (Paris: Editions du Seuil, 1976). An anthology on the revolution of the rural world in West Germany, highlighting technological change, is *Agrarmodernisierung und ökologische Folgen. Westfalen vom 18. bis zum 20. Jahrhundert*, ed. Karl Ditt, Rita Gudermann, and Norwich Rüße (Paderborn: Schöningh, 2001). Case studies concerning the environment and modern agriculture are included in *Ecological Relations in Historical Times: Human Impact and Adaptation*, ed. Robin A. Butlin and Neil Roberts (Oxford: Blackwell, 1995). Modern agriculture and agrochemistry in Scandinavia are covered in Erland Mårald and Jordens Kretslopp, *Lantbruket, staden och den kemiska vetenskapen 1840–1910* (Umeå, Sweden: Institutionen för historiska studier, 2000).

An authoritative work on the subsistence crisis of 1816–1819 is John D. Post's *The Last Great Subsistence Crisis in the Western World* (Baltimore: Johns Hopkins University Press, 1977).

The key source covering Saami culture and use of natural resources is *Law and the Governance of Renewable Resources: Studies from Northern Europe and Africa*, ed. Erling Berge and Nils Christian Stenseth (Oakland, CA: ICS Press Institute of Contemporary Studies, 1998). This work also treats the regulation of fisheries in the Barents Sea. More detail on the Norwegian herring fisheries can be found in Arnvid Nedkvitne, "Kystøkologi og historie," in *Människan och miljön* (see above).

The evolution of northern European woodlands is well treated in the articles by O. Rackham, G.H.P. Dirkx, Matthias Bürgi, G. Tack, and M. Hermy in

The Ecological History of European Forests, ed. Keith J. Kirby and Charles Watkins (Wallingford, UK, and New York: CAB International, 1998). A. S. Mather and J. Fairbairn elucidate the concept of forest transition in "From Floods to Reforestation: The Forest Transition in Switzerland," *Environment and History* 6 (2000): 399–421. See also Gerhard Weiss, "Mountain Forest Policy in Austria: A Historical Policy Analysis on Regulating a Natural Resource," *Environment and History* 7 (2001): 335–355. An analysis of European woodlands and the ecological changes wrought by modern forestry from a resource-management perspective is F.W.M. Vera in *Grazing Ecology and Forest History* (Wallingford, UK, and New York: CABI Publishing, 2000). Tamara L. Whited treats alpine restoration and its reception in France in *Forests and Peasant Politics in Modern France* (New Haven, CT: Yale University Press, 2000).

Early conservation and nature preservation are the subject of a weighty literature that this essay can only begin to touch upon. In addition to volumes by Walter and Delort, and Winter (see above), we recommend David Evans, *A History of Nature Conservation in Britain* (London and New York: Routledge, 1992); Robert A. Lambert, *Contested Mountains: Nature, Development and Environment in the Cairngorms Region of Scotland, 1880–1980* (Cambridge, UK: White Horse Press, 2001); Raymond Dominick, *The Environmental Movement in Germany: Prophets and Pioneers 1871–1971* (Bloomington: Indiana University Press, 1992), the first comprehensive book on the German conservation and homeland protection movements; Willi Oberkrome, *Teure Heimat. Nationale Konzeption und regionale Praxis von Naturschutz, Landesgestaltung und landschaftlicher Kulturpolitik. Westfalen-Lippe und Thüringen 1900 bis 1960* (Paderborn: Schöningh 2004), a history of the German homeland protection movement showing the importance of *völkisch* ideas and ideological continuity from the 1920s to the 1950s; Franz-Josef Brüggemeier, *Tschernobyl, 26. April 1986. Die ökologische Herausforderung* (München: DTV, 1998), a concise history of ecological problems and environmental protection in Germany since 1800; and Patrick Matagne, *Aux Origines de l'écologie: les naturalistes en France de 1800 à 1914* (Paris: Editions du CTHS, 1999), on the fruitful work of nineteenth-century French naturalists. See also Henny J. van der Windt, "The Rise of the Nature Conservation Movement and the Role of the State: The Case of the Netherlands, 1860–1955," in *Yearbook of European Administrative History* (Baden-Baden: Nomos Verlagsgesellschaft, 1999), 227–251; and *Nature in Ireland: A Scientific and Cultural History*, ed. John Wilson Foster (Dublin: Lilliput Press, 1997). Agriculture, forests, water, conservation, and recreation all receive treatment in the engaging essays by T. C. Smout in *Nature Contested: Environmental History in Scotland and Northern England since 1600* (Edinburgh: Edinburgh University Press, 2000).

A rare study of ecological restoration following the ravages of modern warfare is Hugh Clout's *After the Ruins: Restoring the Countryside of Northern France after the Great War* (Exeter: University of Exeter Press, 1996).

Many of the more synthetic studies already referred to carry their analyses well into the twentieth century. A global history of the twentieth century with much information relevant to Europe is J. R. McNeill, *Something New under the Sun: An Environmental History of the Twentieth-Century World* (New York and London: W. W. Norton, 2000). John Sheail's *An Environmental History of Twentieth-Century Britain* (New York: Palgrave, 2002) takes a decidedly optimistic view of efforts to slow ecological degradation in Britain. In chap. 8 of *World Ecological Degradation* (see above), Sing C. Chew places Europe's relations with its economic peripheries in the center of a sweeping analysis of global ecological decline.

A plethora of political and economic histories of postwar Europe have long contributed to our understanding of the processes and contexts that shaped the era's tremendous economic growth, but environmental perspectives are rare. A useful look at postwar consumer society is Christian Pfister's essay, "The 'Syndrome of the 1950s' in Switzerland: Cheap Energy, Mass Consumption, and the Environment," in *Getting and Spending: European and American Consumer Societies in the Twentieth Century*, ed. Susan Strasser, Charles McGovern, and Matthias Judt (Cambridge, UK: Cambridge University Press, 1998). Readers should also turn to Brüggemeier's *Tschernobyl* (see above).

Recent articles in the journal *Environmental Politics* trace the ongoing evolution of Green politics and environmentalism in Europe. On the formation of Green parties see *The Green Challenge: The Development of Green Parties in Europe*, ed. Dick Richardson and Chris Rootes (London and New York: Routledge, 1995). Covering a wide range of topics, Michael Bess treats the partial and paradoxical "environmentalization" of French society in his incisive book *The Light-Green Society: Ecology and Technological Modernity in France, 1960–2000* (Chicago: University of Chicago Press, 2003), a study relevant to all advanced industrial societies. The philosophy and sociology of French environmentalists are treated, respectively, in Kerry H. Whiteside, *Divided Natures: French Contributions to Political Ecology* (Cambridge, MA: MIT Press, 2002) and Jean-Luc Bennahmias and Agnès Roche, *Des Verts de toutes les couleurs: Histoire et sociologie du mouvement écolo* (Paris: Albin Michel, 1992). Kai Hünemörder, *Die Frühgeschichte der globalen Umweltkrise und die Formierung der deutschen Umweltpolitik (1950–1973)* (Stuttgart: Steiner, 2004) looks at early West German environmental politics, showing the impact of both national traditions and international public debates on its emergence. *Konflikte, Konzepte, Kompetenzen. Beiträge zur Geschichte des Natur- und*

Umweltschutzes seit 1945, ed. Franz-Josef Brüggemeier and Jens Ivo Engels (Frankfurt: Campus, 2004), presents the latest German historical research on the politics of environmental protection in the latter half of the twentieth century. With particular treatment of Sweden, Denmark, and the United States, Andrew Jamison places environmentalism in a broad framework in *The Making of Green Knowledge: Environmental Politics and Cultural Transformation* (Cambridge, UK: Cambridge University Press, 2001). Readers can keep up with the latest environmental directives from the European Union by reading the press releases posted at. http://europa.eu.int.

CASE STUDY: FUEL RESOURCES AND WASTELANDS IN THE NETHERLANDS AROUND 1800

Many studies have been published about peat extraction, polders, and the Dutch economy in the Golden Age. See J. W. de Zeeuw, "Peat and the Dutch Golden Age: The Historical Meaning of Energy Attainability," *A. A. G. Bijdragen* 21 (1978): 3–31; J. de Vries and A. van der Woude, *The First Modern Economy: Success, Failure, and Perseverance of the Dutch Economy, 1500-1815* (Cambridge, UK, and New York: Cambridge University Press, 1997); and P. van Dam, "Sinking Peat Bogs: Environmental Change in Holland, 1350–1550," *Environmental History* 5, no. 4 (2000): 32–45. More information on the making of polders during the Dutch Golden Age can be found in John F. Richards, *The Unending Frontier* (see above). Dutch publications used for this case study are J. Trouw, *De Nederlandse veenplassen* (Amsterdam: Callert de Lange, 1948); S. W. Verstegen, *Gewestelijke Financiën,* Utrecht 1579–1798 (The Hague: Institute for Dutch History, forthcoming); and M.A.W. Gerding, "Vier eeuwen turfwinning," *A. A. G. Bijdragen* 35 (1995). This last study contains an English summary on 363–372. Deforestation and afforestation in the northern Netherlands are studied in the two volumes written by J. Buis, "Historia Forestis. Nederlandse bosgeschiedenis," *A. A. G. Bijdragen* 26, 27 (1985). This study has a German summary on 921–924. A synthetic, quantitative article on the sand dunes of the Veluwe is J. Daams, "Van geteisterd landschap tot gekoesterd natuurschoon," *Bijdragen en Mededelingen Gelre* 86 (1995): 116–133. See also S. van Brienen and S. W. Verstegen, "Herbebossing op de Noordwestelijke Veluwe in de 19e eeuw," *Jaarboekje voor de geschiedenis en de archeologie van de Veluwe* 9, no. 10 (2000–2001): 48–64. [This article is published in a single volume of this yearbook, which published the issues of the years 2000 and 2001 together.]

CASE STUDY: THE GERMAN GREEN PARTY

The German Greens figure prominently in the writings of political and social scientists. Historians, however, are just beginning to explore them. The most recent, concise history of the German Greens, centering on their electorate and highlighting their role in the government, is Markus Klein and Jürgen Falter, *Der lange Weg der Grünen. Eine Partei zwischen Protest und Regierung* (München: Beck, 2003). The standard book on the party remains, however, the seminal study by Joachim Raschke, *Die Grünen. Wie sie wurden, was sie sind* (Köln: Bund, 1993). It covers the early history, internal structure, ideology, political strategy, social structure, internal conflicts, and other information on the Green Party. One of the most striking aspects of the Green political culture, grassroots ideology, is analyzed by Dieter Salomon, *Grüne Theorie und graue Wirklichkeit. Die Grünen und die Basisdemokratie* (Freiburg: Arnold-Bergstraesser-Institut, 1992), a study that also examines the practical consequences of this ideology. A best-selling primary source written by one of the Greens' founders is an apocalyptic account of the breakdown of Western societies in the absence of radical change: Herbert Gruhl's *Ein Planet wird geplündert. Die Schreckensbilanz unserer Politik* (Frankfurt: S. Fischer, 1975), which typified public debate in Germany during the 1970s.

CASE STUDY: FROM NATURE CONSERVATION TO SUSTAINABLE DEVELOPMENT: THE SCANDINAVIAN EXPERIENCE

Readers can usefully employ Raymond Dominick's *The Environmental Movement in Germany* (see above) as a point of departure for the study of nature conservation in Scandinavia. Recommended reading about the history of nature conservation in Norway from its early phase up to the present is Bredo Berntsen, *Grønne Linjer. Natur- og miljøvernets historie i Norge* (Oslo: Grøndahl og Dreyers Forlag AS, 1994). A parallel text treating Danish nature conservation is Bent Jensen, *Miljøproblemer og velfærd* (København: Spektrum, 1996). On the debate that led to the early creation of national parks in Sweden, see Sverker Sörlin, *Framtidslandet: debatten om Norrland och naturresurserna under det industriella genombrottet* (Stockholm: Carlsson, 1998).

Hilde Ibsen treats the postwar genesis of environmental politics and a broad consciousness of ecological crisis in *Menneskets fotavtrykk. En økologisk verdenshistorie* (Oslo: Tano Aschehough, 1997), as do Sverker Sörlin and

Anders Öckerman, *Jorden en ö. En global miljöhistoria* (Stockholm: Natur och kultur, 1998).

Nature conservation and environmental protection became part of the concept of sustainable development after 1987. There is already a weighty literature on this topic; a scholarly account discussing sustainable development in Scandinavia is *Implementing Sustainable Development: Strategies and Initiatives in High Consumption Societies*, ed. William M. Lafferty and James Meadowcroft (Oxford: Oxford University Press, 2000). See in particular Katarina Eckerberg's article describing the implementation of goals and visions in Sweden between 1987 and 1999, "Sweden: Progression Despite Recession," and Oluf Langhelle, "Norway: Reluctantly Carrying the Torch." See also Jonas Anshelm, *Social Demokraterna och miljöfrågan. En studie av framstegstankens paradoxer*, Brutus Östlings Bokörlag Symposion (Stockholm: Stehag, 1995), for an important overview and analysis of the Social Democrats in Sweden and their environmental ideas and policies from 1960 to 1995. A key primary source remains the seminal publication by the Brundtland Commission, *Our Common Future: World Commission on Environment and Development* (Oxford: Oxford University Press, 1987).

INDEX

ABOUT THE AUTHORS

Tamara L. Whited, Ph.D., is associate professor of history and assistant chair of the History Department at Indiana University of Pennsylvania, Indiana, PA. Her published works include *Forests and Peasant Politics in Modern France*.

Jens I. Engels, Ph.D., is assistant professor in the History Department at Freiburg University, Freiburg, Germany. He won the Gerhard-Ritter Award in 1999 for his doctoral thesis and is presently writing a detailed study of the transformation of the conservation and environmental protection movements in West Germany from 1950 to 1980.

Richard C. Hoffmann, Ph.D., is professor of history at York University, Toronto, Ontario, Canada. His published works include *Land, Liberties, and Lordship in a Late Medieval Countryside: Agrarian Structures and Change in the Duchy of Wroclaw* and *Fishers' Craft and Lettered Art: Tracts on Fishing from the End of the Middle Ages*.

Hilde Ibsen, Ph.D., is senior lecturer in environmental studies in the Department of Nature and Environment at Karlstads University, Karlstads, Sweden. Her published works include *Visions of Equity: Development Cooperation in Norway, 1980–2000*.

Wybren Verstegen, Ph.D., is lecturer in ecological and economic history at the Free University Amsterdam, Amsterdam, the Netherlands. His many publications include the forthcoming *Gewestelijke Financiën [Provincial Finances]*, *Utrecht 1579–1798*.